HYENA NIGHTS
& KALAHARI DAYS

HYENA NIGHTS
& KALAHARI DAYS

Gus and Margie Mills

JACANA

To the memory of our parents:
Kenneth and Sybil
Harry and D'Urban

First published by
Jacana Media (Pty) Ltd in 2010

10 Orange Street
Sunnyside
Auckland Park 2092
South Africa
+2711 628 3200
www.jacana.co.za

Cover photograph of two cubs
by Lex Hes

ISBN 978-1-77009-811-4

Set in Sabon 10.5/15pt
Job No. 001145
Printed and bound by
CTP Printers, Cape Town

ISO 12647 compliant

See a complete list of Jacana titles at
www.jacana.co.za

Contents

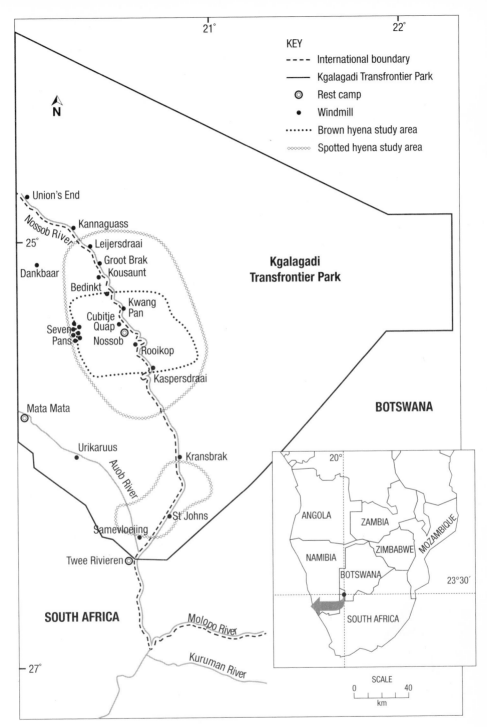

The Kgalagadi Transfrontier Park showing places mentioned in the text and the hyena study areas. During the time of the study, the area was split into the Kalahari Gemsbok National Park, South Africa, and the Gemsbok National Park, Botswana.

Family trees

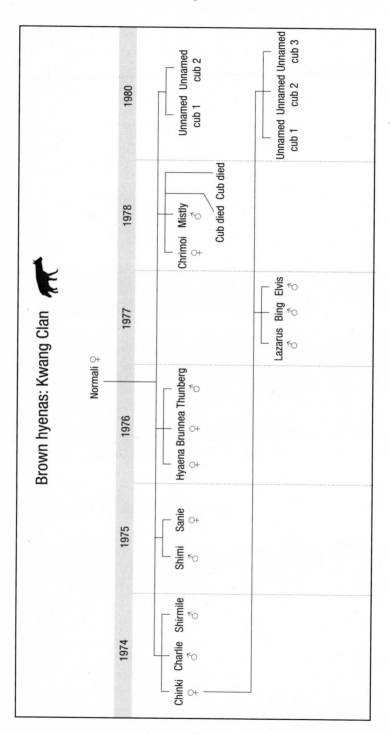

Brown hyenas: Kwang Clan

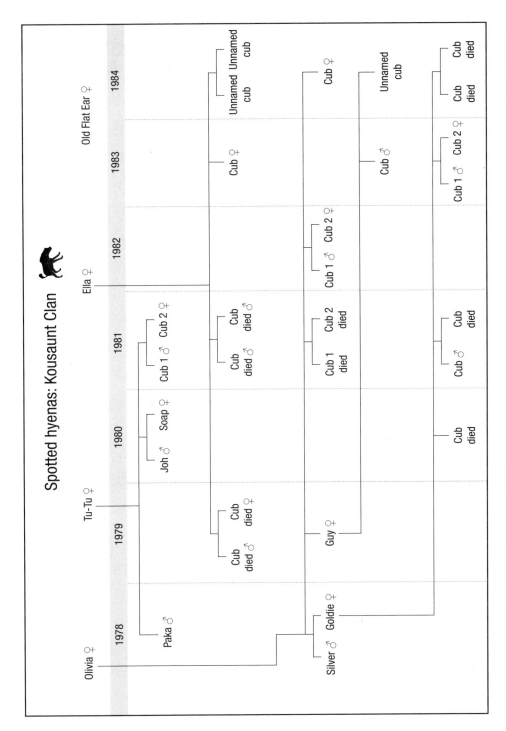

Spotted hyenas: Kousaunt Clan

Prologue

HYENAS HAVE A BAD REPUTATION. They are regarded as ungainly, unintelligent, uninteresting, unattractive and unlovable. In African folklore the hyena is depicted as an abnormal and treacherous animal which embodies magical powers. An article in the conservation magazine *African Wildlife* in the 1960s used words like freak, stupid, demented and ungainly to describe the spotted hyena. It noted that the hyena has been hated and despised down the ages and argued this was only natural because it is such a coward. More recently the Walt Disney film *The Lion King* has reinforced this stereotype. Who cares about these unfortunate animals and who would want to read about them, particularly when there are so many noble and beautiful African carnivores like lions, leopards and cheetahs?

We have had the privilege of being able to spend 12 years together in the Kalahari, one of the most beautiful natural environments on earth, and Gus has spent the best part of the last 30 years studying most of the large carnivores of Africa: lion, cheetah, wild dog, spotted hyena and brown hyena. The cats are certainly cool and the wild dog is wild, but the hyenas are heroes. They are so different from the popular notion of the species. They are intelligent, powerful and, yes, even beautiful: have you ever looked into the big brown eyes of a spotted hyena? Perhaps they are a bit spooky – spotted hyenas can make some amazingly human-like cackles and hysterical laughs – and even at times unpredictable, but that makes them even more fascinating.

As stunning as the Kalahari is, there is a price to pay for living there. The isolation can become difficult to cope with. We lived 420 kilometres from the nearest town and had no contact with the outside world except by weekly mail (if we were lucky) and the official parks radio network. Just to get to a telephone involved a 320-kilometre round trip. The climate is severe – very hot in summer and cold in winter. We had no air conditioners or electric heaters to relieve the conditions, as the generator that provided the power supply to our house was too weak to run such equipment and in any case was usually on for only about six hours per day. With small children these problems were exacerbated.

We have written this book in order to share some of the excitement and emotions we experienced while watching and studying hyenas – the thrill of the chase by spotted hyenas across the dunes after gemsbok at 50 kilometres per hour at night, the relief when a brown hyena mother finally gets back to feed her small cubs at the den after a long night looking for food – events which have taken place every night for thousands of years, yet which we are two of only a handful of people ever to have had the privilege to witness. This aspect is mainly dealt with by Gus in Part 1. Here we hope to educate people about the true nature of the wonderful animals we studied and lived with and to explain the significance of their behaviour in the Kalahari ecosystem. We also hope to change the attitudes of those who have been brought up to believe all the myths and fables about hyenas and are constrained by their prejudice towards these much maligned creatures.

In Part 2 Margie restores the balance and illustrates the problems and pleasures of living in this unusual and challenging environment. She was able to call back the past by reading through the weekly letters that we wrote to our sets of parents. As they were always very detailed we used to carbon-copy them, one week sending the original to one set of parents, the next the copy, and vice versa to the other set. Luckily they had the foresight to keep each and every one and handed them back to us many years later. It has been fun rereading them and thereby reliving the past. We have found it strange how some incidents are so clear that it is as if they happened just yesterday, while others turn out to have been quite different from how we remember them and yet others that we wrote about and that were obviously important at the time we have no recollection of!

It is our desire that people will become more tolerant and respectful of hyenas and indeed all forms of wildlife and that the needless cruelty towards and senseless persecution of what are often innocent victims may stop. We realise that in many areas it is very difficult or impossible for large carnivores and man to coexist, but a brown hyena does not normally kill cattle and spotted hyenas do not ravage impala or suppress lion populations. Merely killing predators is unenlightened and often not the best solution to a problem that may not even exist except in the mind of a land owner or wildlife manager. It surprises us to find how many so-called experts there are when it comes to managing wildlife in general and predators in particular. The relationship between predator and prey is as intricate and complex as any ecological relationship. At a time when issues like overpopulation (the real inconvenient truth) and global warming are seriously undermining the future of the planet, we need to change our mindset and be more innovative, thinking less about our own material well-being and more about sharing resources, not only among our own species but among all forms of life.

Acknowledgements

MANY FRIENDS, family and colleagues gave us much help and encouragement during our 12 years in the Kalahari (1972 – 1984) and latterly while writing this book. It is a pleasure to acknowledge their contributions and to thank them for their help.

Professors Koos Bothma and Fritz Eloff, from the University of Pretoria, were instrumental in setting up the original brown hyena project in the Kalahari and paving the way for us to get there. The National Parks Board (now called SANParks) and the Department of Wildlife and National Parks, Botswana, gave us permission to work there. As Gus was an employee of the National Parks Board for all but our first two years, the major funding for the project was through the board, with additional funding from a number of conservation organisations: the SA Nature Foundation, the Wildlife Society of Southern Africa and the Endangered Wildlife Trust.

Hans Kruuk, Martyn Gorman and David Macdonald provided much valuable scientific advice and guidance and also the encouragement to keep us going, especially during the inevitable difficult times. At Nossob, a succession of rangers Elias le Riche, Div de Villiers, and Rian Labuschagne provided valuable logistic support. Our trackers, Ooi and Houtop for the brown hyenas, and Hermanus Jaggers for the spotted hyenas, were invaluable.

We made a number of good friends during our time in the Kalahari and they helped us in numerous ways, not least in providing much-needed social opportunities in such an isolated area. Clem Haagner and Richard Liversidge were frequent visitors and always arrived laden with goodies to eat and drink. Richard, in particular, always gave us guidance and advice in helping to overcome any challenging issues we might have been dealing with at that time. Richard and Karen Goss and Rian and Lorna Labuschagne lived for several years at Nossob and we had lots of fun together.

Our parents (Gus's, Kenneth and Sybil, and Margie's, Harry and D'Urban Davies) were a constant source of support. Not only did they visit us frequently, but if ever we needed anything (and we often did) they were always there to lend a hand. Gus's parents, living in Johannesburg, were especially well placed to help and always did more than we asked. We are also very grateful to them for keeping all our letters from which we were able to recall many incidents reported here. Once we had children, Margie's Aunt Con and Uncle Bob were often wonderful babysitters allowing us to get on with other tasks.

Ian Michler and Val Thomas were the catalysts for getting us to finally complete this work. Ian for suggesting we contact Val, and Val for paving the way to our publishers. Several friends were kind enough to comment on our early drafts: Pam Blundell, Paul and Katinka du Preez, Roxanne Reid, and Margie's sister, Alison McLean. Our editor at Jacana Media, Russell Martin, made many helpful comments and suggestions, and Shawn Paiken was responsible for the layout and design.

PART 1

Hyena Nights

Gus Mills

CHAPTER 1

Getting to know Brown Hyenas and the Kalahari

I WOKE AT SUNRISE. I looked out of my caravan window onto a vista of coppery red sand, peppered with grass tussocks and dotted with short, stumpy trees. It was my first view of the Kalahari and, although I had seen many photographs of the landscape, no photo can convey the beauty and atmosphere of this region at sunrise. I immediately knew that I would be at home here.

It was April 1972 and I was on my way to the Kalahari Gemsbok National Park to start a two-year study of the little-known brown hyena. I had left Johannesburg the previous day in my brand-new Land Rover pulling a three-metre caravan, which was to be my home for the next two years. Actually it was to be our home, as in a couple of months I was due to marry Margie Davies, my university sweetheart, who was recovering from knee operations in her home in Salisbury, Rhodesia (now known as Harare in Zimbabwe).

I had spent the night on the side of the road that runs along the Kuruman river-bed from Kuruman to Twee Rivieren, the southern entrance to and headquarters of the Kalahari Gemsbok National Park. After a quick cup of coffee I was soon on the road again, feeling excited but also apprehensive about the great adventure that lay ahead of me. Even in my wildest dreams that morning it did not cross my mind that I would spend the next 12 years in the Kalahari, nor did I have any idea that I would witness such a multitude of incredible and exciting natural history events.

3

Ever since I first visited the Kruger National Park as an eight-year-old I had wanted to work in the bush with animals. When I left school with a less than adequate academic record, I was advised not to try for a science degree. Instead I registered at the University of Cape Town for a BA in psychology. After three years, during which time I became even more convinced that I wanted to study animals and not people, I emerged with a very mediocre degree. The prospect of becoming a personnel officer or clinical psychologist was unacceptable.

My parents, Ken and Sybil, who were always so supportive, gave me a second chance and I went back to register for a BSc in zoology. Since then I have never looked back. In 1971 I registered at Pretoria University for an honours degree in wildlife management. During this year I heard from my professor, Koos Bothma, that he wanted a student to conduct a basic ecological study of the brown hyena in the Kalahari. Although until then hyenas had not been animals that I had thought particularly interesting, the prospect of going to live and work in the Kalahari for two years was most appealing.

Accordingly I set about trying to raise funds for the project. After several frustrations and disappointments I managed, through the help of Professor Fritz Eloff, head of the Department of Zoology at Pretoria, Professor Bothma and a friend, Mark Berry, to raise enough money to start the project. I also managed to gain permission from the National Parks Board of South Africa and the Department of Wildlife and National Parks, Botswana, to live and work in the Kalahari Gemsbok and Gemsbok National Parks in the southern Kalahari.

For the moment, however, I had other things to concentrate on. Two kilometres north of Andriesvale, where the Kuruman river-bed runs into the Nossob and the road turns north, I came to a large stretch of water in the dirt road, easily 200 metres long, the result of an unusually heavy downpour a few days earlier. I stopped. While I was wondering how to get through this obstacle, a woman driving an old Peugeot 404 passed me, drove straight into the middle of the water and out on the other side. If she can get through this in her old saloon car, I thought, then I should have no trouble in my new Land Rover, even though I am pulling a caravan.

I put my vehicle into four-wheel drive, just in case, and in I went. Gradually I felt the vehicle labouring more and more, until, not quite halfway through the pool, I ground to a halt. In retrospect I realised that the mistake I had made was to go to the side of the road and not to stay in the centre as the woman in the Peugeot had done. Because of the camber in the road, the sides were far deeper. I unhitched the caravan and managed to drive the Land Rover out of the mud. Then with the help of a few local children (it's amazing that in so many parts of Africa within minutes of a car stopping people emerge from the bush to see what is going on!) we managed to pull the caravan on to the higher ground.

I rehitched the caravan and off I went. However, something was still wrong as the Land Rover was struggling to pull the caravan. I stopped to inspect and found that I had failed to lift up the little front wheel of the caravan that stabilises it when it is not attached to a vehicle. Now I was in trouble, as the pipe attaching the wheel to the Jurgens had become bent, from being forced through the mud, and I couldn't pull it up. With the help of my new friends we managed to lift the front of the caravan out of the mud and to extract the wheel with its bent pipe from below. Fortunately the garage at Andriesvale was open and the mechanic soon managed to straighten the metal for me. Within half an hour I was on my way to Twee Rivieren again, having learnt a few valuable lessons on how to negotiate the infamous roads of the Northern Cape.

As I approached Twee Rivieren the road deteriorated. The rain pools in the road became longer and deeper. By staying in the middle of the road when negotiating the pools and by following the tracks of other vehicles when they deviated from the road for a particularly large pool, I managed to reach Twee Rivieren with no further mishap, or so I thought.

When I opened the caravan it looked like a brown hyena had been at work inside. Because of the bumpy ride over the deviations, the mirror on the little clothes cupboard had fallen off and put a hole through the cupboard door. The sun-roof had also fallen off and the entire surface area of the caravan was covered in dust. To crown it all, a five-litre can of cooking oil had sprung a leak, so that the floor was also covered in a thin layer of cooking oil. I spent the best part of my first two days in the Kalahari cleaning and repairing my caravan.

The Kalahari Gemsbok National Park in the Northern Cape Province of South Africa and its neighbour, the Gemsbok National Park in Botswana, form one of the largest national park systems in the world, covering 36 000 square kilometres. Recently the two parks have been merged into the Kgalagadi Transfrontier Park through an international agreement. An arid region receiving on average 220 millimetres of rain per year, it also experiences great temperature fluctuations. In summer the mercury usually climbs to over 35°C and on cold winter nights it can plummet to –10°C.

The centres of activity are the two fossil river-beds, the Auob and the Nossob, which enter from South West Africa and run roughly north-west to south-east through the Kalahari Gemsbok National Park. The Nossob forms the boundary between South Africa and Botswana.

It had been decided that I was going to make Nossob Camp my base, so after spending a few days orientating myself at Twee Rivieren, I moved up the Nossob river-bed to the camp 160 kilometres away. At that stage Nossob had only eight huts for tourists and a small camping ground, and was run by the ranger Elias le Riche, Kalahari-born, and his wife Doempie.

Obviously I was anxious to see my first brown hyena. Every morning I was out at sunrise and back after dark, driving slowly along the Nossob River valley and over the dunes, hoping to catch a glimpse of my elusive study animal. From talking to Elias and others who were familiar with the area, it soon became obvious that these animals were nocturnal and difficult to contact. I knew that this would be the case, but no one had told me it was going to be so difficult just to see the animal!

An outstanding feature of the Kalahari is the layer of red windblown sand that covers over 95 per cent of the area. Not only does this contribute to the beauty of the area, it also means that it is possible to follow tracks of animals in the sand, something that the local San people excel at. This sand takes on a variety of colours, ranging from vivid orange-reds to burnt amber, from brown ochre to smoky-pink. The red hue is caused by a covering of natural ferric iron oxide on the minute sand grains. Because of the low rainfall, moisture has not been able to oxidise the ferric oxide to white ferrous oxide.

Early one morning a Bushman tracker, Ooi, and I set out following the spoor of a brown hyena that had visited the rubbish tip at the camp the previous night. Was this to be the day that I caught a glimpse of my first brown hyena? The spoor went into the Nossob river-bed and then disappeared. At this early stage I discovered one of the great limitations to the technique of tracking in this area. The hard substrate makes it impossible to follow an animal once it moves into a river-bed. Unless the animal happens to cross straight over, and its tracks can be picked up again on the other side, it is invariably lost. This was most unfortunate, as there was always plenty of food for hyenas along the river-beds, and they spend much time foraging there. From tracking spoor we would not be able to learn what they were doing in this important habitat.

Ooi spent the next two hours searching the sides of the river-bed for the tracks to emerge, but to no avail. Frustrated and disappointed, we returned to Nossob Camp. The next morning Ooi found another set of spoor at the refuse tip. Again this led us straight into the river-bed and my heart sank. However, when he emerged on the opposite bank and went up the dune into Botswana, my spirits rose. This time the hyena had crossed straight over. We followed the tracks across the dunes, Ooi walking while I followed some way behind in the Land Rover. Progress was slow as it had moved erratically. At one point we spent over an hour progressing only 100 metres, following a series of repeated circles. Then the hyena started moving forward again. No clues were left as to what the animal had been doing.

About one kilometre further on Ooi stopped to bend down and pick something up. I drove up to find him holding the shell of a melon-like fruit. On looking around I discovered we were in a large patch of tsama melons, the wonder plants of the Kalahari, which consist of over 90 per cent water. Some had been freshly opened and the flesh neatly scraped out, while others were dried-out shells that had obviously been eaten some time ago. There were still others that were intact and had not been eaten at all. The brown hyena had obviously enjoyed munching on several of them. This was my first brown hyena feeding record and I was somewhat surprised that its meal should be a fruit. Little did I realise what an important role these melons play in the diet of both animals and humans in the Kalahari.

Feeling excited by our first discovery, we continued on our way. It was nearing midday and with the sun high in the sky it became more difficult to see the tracks because of the lack of shadow, which enhances the outline and therefore the clarity of the spoor. I was amazed by Ooi's ability. Several times I asked him to show me the spoor, but all I could see was a slight indentation in the sand, which looked no different from many other imprints. Over the years I never ceased to be amazed at the skill of the local people in following an animal's trail. Although as time went on I myself improved in tracking competence, I was never able to follow the spoor of a single hyena when the sun was high with the same expertise as Ooi and his people.

Matching this uncanny aptitude, the Kalahari trackers have incredible stamina. They are able to trot over the dunes for hours on end in high temperatures. Since leaving the Kalahari I have become an avid jogger myself and know what can be achieved by someone who is fit and in training. Bushmen, however, are natural athletes and are able to keep moving for 20 kilometres without suffering any effects, even if they have not run for several months. This is in soft sand, over dunes and without any fancy running shoes or energy-boosting supplements.

Suddenly Ooi stopped and pointed ahead of him. About 10 metres away, a rather ungainly large, dark bundle of hair, supported by four thin legs, was running away from him. He had flushed the brown hyena from a small bush under which it had been sleeping. After a short distance the hyena stopped and looked back. I caught a glimpse of its large pointed ears and a white ruff around the neck. Then it turned round and with long black coat flowing in the wind, like a schoolmaster in academic gown hurrying to class, it disappeared over a small dune. It was a thrilling moment, for at last I had made contact, however limited, with my study animal.

I spent a month in the Kalahari, familiarising myself with the area and the conditions under which Margie and I were going to work and live before journeying up to Salisbury for our wedding. We must be the only married couple who had two hyenas on top of their wedding cake in place of the more usual bride and groom.

When we returned to the Kalahari I began to plan my research strategy seriously. The brown hyena is a member of a rather exclusive family of carnivores, the Hyaenidae, which has only four living members. There are the three true hyenas – the striped hyena (*Hyaena hyaena*), the brown hyena (*Parahyaena brunnea*) and the spotted hyena (*Crocuta crocuta*) – and a rather aberrant member, the aardwolf (*Proteles cristatus*).

As the brown hyena was such a little-known species, the main objective of the study was to collect as much information on the animal's ecology and behaviour as possible. The first question to ask seemed to be, what do they eat and how do they find their food? At the same time we wanted to know how they move and what the size of an individual's home range is and what kind of social system they have. Other important questions were: how do they raise their young, how do they communicate and how do they interact with other carnivores?

What we did know was that the true hyenas are characterised by having large and robust jaws and teeth, useful for crushing bones. On the other hand the aardwolf has a few weak peg-like cheek teeth, although its canines are quite well developed. The aardwolf is one of the most specialised of carnivores, feeding almost exclusively on harvester termites, for which teeth are not a priority. Although it is so different from the true hyenas, it is included in the family on account of close anatomical, chromosomal and blood protein relationships. Furthermore, in common with the true hyenas, it possesses an anal pouch. This unique organ, used for scent-marking (about which I will have more to say later), is situated between the rectum and the base of the tail and can be turned inside out. From it the animals deposit a strong-smelling paste-like substance on to grass stalks by carefully stepping forward over a stalk and placing the extruded anal pouch on to it. No other animals scent-mark in this way.

Surprisingly hyenas are more closely related to cats than to dogs. Their nearest relatives, in fact, are the cat-like civets and genets of the family Viverridae. Hyenas are believed to have evolved in Africa and Eurasia from a civet-like ancestor some 26 million years ago. The early hyenas did not, as it would be logical to assume, develop large teeth and massive bone-crushing jaws. They actually looked more like modern-day dogs. The true dogs of the family Canidae were evolving in North America, so this dog-like niche was available to the hyenas in Eurasia and Africa. When the canids were able to cross the Bering land-bridge into Eurasia, the dog-like hyenas began to disappear, although palaeontologists do not think that these two occurrences were necessarily related. The only dog-like hyena that managed to survive was the ancestral aardwolf.

The large bone-crushing teeth in hyenas evolved during the late Miocene era, 10 million years ago. It was then that the hyenas reached their peak in terms of numbers of species, there being 10 living together, including one as large as a cow. It is thought that during this time an abundant supply of carrion was available, as the huge but highly specialised sabre-toothed cats could not deal with the tougher portions of their large prey like woolly rhinoceroses. As the sabre-toothed cats declined during the early Pleistocene about two million years ago, so did the hyenas.

Margie's oldest brother and his wife, Rob and Roz, had given us a copy of Hans Kruuk's classic work *The Spotted Hyena* for a wedding present. It was this work that first showed that spotted hyenas are not only scavengers, but efficient hunters in their own right, taking large prey like wildebeest and zebra on a regular basis. Reports from the Kruger National Park suggested that there the brown hyena, although rare, is an efficient hunter and that the common spotted hyena is mainly a scavenger. As it turned out, these reports were not accurate, but at that time we were anxious to find out how much hunting brown hyenas did in the Kalahari.

It was also important for us to establish sound and reliable methods for collecting the type of information we were interested in. We had always envisaged that tracking spoor was going to be the major method.

Professor Fritz Eloff had for years used this method successfully for studying lions in the Kalahari. However, as I have already mentioned and as we were soon to learn, there are limitations to this technique with brown hyenas. Although we did make extensive use of it in the early months, we eventually replaced this technique with the far more valuable, not to say exciting, method of directly following the animals and observing their behaviour at first hand.

For the moment, however, tracking spoor occupied most of our lives. I had now managed to employ the full-time services of Ooi's brother, Houtop, as Ooi was often needed to fulfil his duties as a ranger. Every morning at sunrise the three of us were out searching around the camp or at one of the windmills close by for brown hyena spoor. When we were lucky enough to find one, we would follow the spoor for as long as we could. Occasionally we would catch up with the animal and were rewarded with a fleeting glimpse of it as it ran off, having been flushed from its daytime resting place. In spite of chalking up several hundred kilometres of brown hyena spoor, we obtained no evidence for their hunting anything. All spoor we found were of single animals and the only items we recorded them as eating were tsama melons and gemsbok cucumbers, a similar type of wild fruit, as well as the bony skeletons of antelope such as springbok and gemsbok.

We also needed to expand our observations of the brown hyena's feeding habits. A well-tried and trusted method in carnivore feeding studies is through faecal, or scat, analysis. So we started to collect scats in order to analyse these for the remains of food items. This was a monumental task for which Margie takes the credit. Over the next two years she analysed 383 scats from which she identified 1389 food items. These included an assortment of different types of food from hartebeest to honey badgers, snakes to suricates and bat-eared foxes to beetles. While small and large mammals, beetles and wild fruits, made up the majority of food types, birds, reptiles, birds' eggs and other insects were eaten as well. The brown hyena, we were to learn, has a catholic diet.

Although we were slowly beginning to get an idea of the diet of the brown hyena, we were learning little about their social behaviour, except that they always travelled alone, as we were unable to identify any individuals. Nor had we learnt anything about their breeding habits or found any cubs. It was slow progress for much effort. We were beginning to learn an important lesson in the study of animal behaviour, patience and tenacity (the famous words of Churchill often kept us going 'Never give up, never, never!') and a lucky break was just around the corner.

CHAPTER 2

A Roller-Coaster
Ride

ONE MORNING IN JULY we were out early looking for spoor at Cubitje Quap Windmill a few kilometres up the Nossob river-bed from Nossob Camp. Houtop soon found one and followed it into the dunes. We were pleased that the animal had moved away from the river-bed. After we had been going for an hour, during which time we moved about four kilometres, Houtop stopped at a large mound of sand. He beckoned to me to come closer. As I approached the mound I realised that it was the entrance to quite a large hole, going down at an angle of about 45° and quickly tapering down to a narrow tunnel about 40 cm high and 60 cm wide. Scattered around the hole were several skulls and bones, one or two springbok horns, as well as several hyena scats. We had finally found a den with cubs, something we had been hoping to do for some time. Now, perhaps, we could make some behavioural observations of the hyenas.

With our hopes higher than they had been for some time, we returned to the den that evening. Fortunately there was a high dune about 400 metres away, giving us a good view of the den and its environs. We took up our position and, after about an hour, just before it became too dark to see from this distance, we observed a brown hyena approaching the den. It sniffed around the mound in front of the den where we had been walking several hours earlier, and then lay down in a small hollow next to it. From the distance and in the poor light we were unable to tell whether it was an adult or a large cub. In Africa the dark deepens rapidly, and unwilling but excited, we returned home, as it was pointless staying any longer.

We set out before sunrise the next morning, armed with a flask of coffee and blankets to fight off the sub-zero temperature, and again took up our position on top of the dune. When it became light enough to see, we could make out two brown hyenas lying curled up next to each other outside the den entrance. After some time one of them stood up, sniffed around the opening and then disappeared into the hole. Then the second one also stood up and moved off to the right. We agreed that both animals looked like large cubs, smaller and less robust than an adult.

For the next week we visited the den each sunrise and sunset and only ever saw the two cubs. The pattern was similar to our observations on the first morning and evening. The cubs spent most of the time lying together in the little hollow next to the entrance of the hole. Sometimes both would disappear into the den for the day, and sometimes either one or both would move away, returning in the evening. We later learnt that they spent the day in another hole close by or else slept under a bush. We were disappointed that we had not seen any signs of adult hyenas with them. They certainly did not seem to live there with the cubs.

As full moon was approaching we decided to spend a few complete nights at the site to see if adults were visiting the den in the dark, and to discover how the cubs were being fed. On the first night we took up a position in the Land Rover about 40 metres from the den at 5.30 pm, an hour before sunset. This was a drastic change from our previous observational distance of 400 metres, but we reckoned that there was no use staying up all night if we were unable to see what happened. We reasoned that the stationary vehicle would not disturb the animals. It was a beautiful, clear Kalahari moonlit night. As the sun set and it became darker, our eyes grew accustomed to the conditions and we were surprised how much detail we could see, particularly through our 7x50 binoculars. Soon the first cub emerged from the den. We were rather apprehensive that it might not like us so close and were relieved when, after looking intently at us for several seconds, it lay down in its usual position next to the mound.

Ten minutes later the cub suddenly lifted its head and looked off to its left. Although we strained to see what had alerted it, we could not detect anything. Then gradually in the shadows we could make out the shape of the second cub slowly approaching. Its hair was raised and it did not appear to notice the Land Rover. The first cub stood up and also raised its hair. The two hyenas sniffed each other on the nose briefly, then smelt around the den, occasionally picking up a small scrap of bone in the sand, but largely ignoring each other. After not more than 10 minutes they were both lying down and, after a good grooming session, both went to sleep. It was 7.30 in the evening.

By midnight neither cub had moved and we were feeling rather tired and bored. This was another lesson in patience. Sitting at brown hyena dens certainly brought home to us that in this business patience is not only a virtue, but essential. By sunrise the cubs were still lying in the hollow. Soon they stood up, stretched and moved off in opposite directions.

The next night we were better prepared for the vigil. We brought along some bedding so that one of us could sleep in the back of the truck while the other stayed awake. Margie took first shift from 9 to 12, while I settled down in my sleeping bag. At this stage there were no hyenas in sight. I was awakened just after midnight by Margie tapping on the window and pointing excitedly in the direction of the den. Lifting my binoculars I could see the two cubs circling a large hyena standing on the mound. Whining loudly, with their ears flattened out sideways, they repeatedly cut in front of the adult and presented their hind regions to its face. After sniffing the anal region of the cubs a few times, the adult lay on the mound and the cubs started to suckle. We were surprised to see such large cubs, whom we judged to be a about a year old, still suckling. I was also surprised at how long the mother suckled them for. After 40 minutes she eventually rolled over onto her stomach and the cubs had to stop. She lay there for another five minutes, before disappearing into the night.

We were then treated to a marvellous display by the cubs of brown hyenas at play in the moonlight. Initially they stood on the mound of the den, face to face, then suddenly dropped onto their elbows and attempted to bite each other on the jowls and the side of the neck. I had noticed that just below each ear brown hyenas have a white spot of smooth hair and this seemed to be a target. They pitched their heads from side to side, parrying and thrusting at each other. One of them managed to grab the other by the side of the neck and to shake it vigorously for a few seconds before losing its grip. At this the second animal bounded off, pursued by the first. They ran around the den, then back onto the mound where the chased animal turned round and faced its pursuer. Then they proceeded to wrestle with each other again. This play, which we called muzzle-wrestling, went on for nearly half an hour, with the roles of attacker and attacked, pursuer and pursued, alternating from one to the other. Eventually, exhausted, the two cubs flopped down next to each other.

A few days later we arrived just after sunrise to find no cubs outside the den. We settled down to wait for them to appear. We hadn't been there five minutes when a large hyena approached with a huge piece of meat in its mouth. It came up to the den, stood on the mound for a few seconds and then put the food down. This was the closest we had ever been to an adult brown hyena. We were so excited we could hardly move. Slowly, without wishing to disturb the hyena, I lifted my binoculars to my eyes and focused on it. In full frame for the first time, I was struck by its power and beauty. It stood proud on the mound of the den looking straight at us with its large pointed ears cocked. I noted the long, stripy legs and powerfully built neck and shoulders. The white ruff on the neck and shoulders was striking against the black shaggy coat. What a handsome beast, I thought; anyone who says brown hyenas are ugly must either be blinded by prejudice or lacking in judgement. After a while the hyena decided it had had a good enough look at us and stuck its head into the den momentarily, before lying down on the mound. Still no cubs appeared. After five minutes the hyena moved away. We waited 10 more minutes and still no cubs appeared. They had obviously moved away from the den before we arrived. I couldn't contain my curiosity any longer to see what it was that the adult had left behind. We approached the den cautiously to find an almost uneaten porcupine with its quills plucked. This was the first evidence we had of adult brown hyenas bringing food back to the den for the cubs. It was also the only time we saw a brown hyena eat a porcupine, although we did find a few quills in some scats. To this day I wonder how the hyena came upon the porcupine and how it had been plucked.

By this time I had obtained some radio transmitter collars from the Council for Scientific and Industrial Research (CSIR) in Pretoria, which had recently started to develop the science of radio telemetry. When we tracked an animal's spoor it was important to know which individual it was we were following, so that we could work out the hyena's movement patterns and territory sizes. I thought that if we could track an animal by radio and find where it was sleeping during the day, we could then return the next day and follow its spoor. But how were we to catch a hyena to fit a radio collar? Elias le Riche had told me that it was impossible to catch a brown hyena in a trap. He suggested that we follow the spoor of an animal to its day-time resting site, then pursue it in the Land Rover as it ran off, firing a dart at it when we were close enough.

The hyena that had brought the porcupine to the den was the ideal recipient of the first collar. As soon as we had examined the den we returned to Nossob Camp to prepare for our first hyena capture. Elias agreed to drive the vehicle and show me how it was done. Back at the den Houtop easily picked up the spoor of the animal. We followed closely in the Land Rover. As it had been at the den after sunrise, the chances were that it would not have moved too far before resting for the day. I felt excited, but also very nervous. Then it happened. We hadn't gone a kilometre when Houtop flushed the hyena. Elias put his foot down and off we sped in pursuit of the fleeing animal. We soon caught up with it running at about 40 kilometres an hour. Elias fired the dart from the Cap-Chur pistol, the dart struck the animal in the back and we left it for the drug to take effect. Ten minutes later Houtop once again took up the spoor and we soon found our quarry sound asleep.

We were surprised to discover that it was a male. (Brown hyenas do not have the same complicated sexual organs as spotted hyenas, the male and female of which are difficult to tell apart.) We assumed that he was the father of the cubs, although as we were to discover later this is not necessarily the case. Being our first hyena in the hand, we examined, measured and photographed him from all angles, before placing the collar around his neck. Now we had to wait for him to recover from the immobilising drug. We had used the drug phencyladine hydrochloride, commercially known as Sernylan, later to become one of the most awful hallucinatory drugs used by addicts, known to them as 'Angel Dust'. I once read of a person who had a bad trip on 'Angel Dust', tearing his eyes out with his fingers. Presumably the drug does not affect animals in this way, but they do take a long time to recover from its effects as there is no antidote. We had also, in retrospect, given the animal rather too heavy a dosage. It was to be 36 hours before he recovered. Margie and I had to stay with him for the entire period, in case he was attacked by other predators. We were very relieved when he eventually walked off more or less in control of himself. We called him Harken after our fathers Harry (Margie's) and Ken (Gus's). We decided that the highest honour we could bestow on anybody was to name a hyena after him.

During the next two weeks we managed to catch and radio-collar the mother of the cubs, who for some obscure reason we called Gertie. We also caught and marked the cubs by cutting small nicks out of their large pointed ears for future recognition. Almost daily we tracked the spoor of one of these animals. The cubs were foraging quite far from the den, sometimes up to 12 kilometres away. Although they were obviously almost independent, they always returned to the den during the night. The two adults only rarely visited the den, the mother arriving on average once every three nights, the male perhaps once a week.

By taking compass readings on Houtop as he disappeared from view when following a spoor, driving along the compass bearing to the point where he had disappeared, and noting the distance travelled from the vehicle's odometer, we were able to map the movements of the hyenas we tracked. In this way I hoped to accumulate enough data on these hyenas to measure the size of their territory and the manner in which they used it. Today this seems such a crude method compared with the GPS equipment now used for this purpose.

Then disaster struck. Within six weeks of fitting the radio collars, both stopped functioning. I was anxious to recover them in order to discover what had gone wrong. Once again we had to employ cowboy tactics to catch the two hyenas by chasing them in the Land Rover over the dunes. Although we cut down the drug dosage, they both still had prolonged reactions to the drug. We sat with the female for over 36 hours. As she was beginning to recover and take her first uncertain steps, she was attacked by six spotted hyenas. She was semi-mobile and ran away when I tried to protect her by manoeuvring the vehicle between her and the attackers. Once the spotted hyenas had brought her down I was able to chase them off, but she had been severely bitten in the stomach. We took her back to camp and I stitched her up, but in the morning she was dead. Our anguish increased when we discovered that she had an almost full-term foetus inside her.

Then it was Harken's turn. Although he had apparently recovered from the drug and was moving quite well, he drowned in a drinking trough the day after we caught him. Subsequently I have heard of two other animals, a badger and another brown hyena, that drowned while recovering from Sernylan. To add insult to injury the new collar that we had put on the male worked only for a few hours. We were devastated. This was certainly not what we had envisaged when we set out on this venture. We were close to packing up and going home. Not for the last time Churchillian philosophy kept us going!

We were in a desperate situation. The radio collars were clearly not performing as efficiently as we wanted them to. Even more serious, we would have to change our capture technique. It was far too traumatic for both the animals and us. Having been working virtually full-time on the project for nearly six months, we decided to take a break and went up to Pretoria for a couple of weeks. I needed to discuss my project with my supervisor and visit the CSIR to see if they could do something to improve the radio collars. We also needed to relax a bit with friends and family. Unfortunately we were never to have much success with the locally made collars and it was only when we used imported collars from the US that we were able to get meaningful results with radio telemetry.

On our return to the Kalahari we managed to track the spoor of the cubs a few times and catch a glimpse of them running off. However, we needed to make contact with an established breeding female and so our priority became to find another den. Fortunately it wasn't long before our luck turned. We followed a spoor from Kaspersdraai, about 20 kilometres south of Nossob Camp, into Botswana. Some 15 kilometres from the river-bed we came across a much smaller hole than the previous den, which, judging by the size of the spoor around it, was the home of small cubs. There was no high dune in the vicinity from which to make observations, so we were forced to park the Land Rover about 30 metres from the den. Just before sunset we saw an adult brown hyena approaching from the opposite side. She stopped when she was about 10 metres away and looked at the vehicle intently. We held our breath, not daring to move a muscle. Then she moved closer and we breathed a collective sigh of relief. She came up to the hole, put her head inside for a few seconds, and then lay down on the mound. Three small cubs, exact replicas of the adults except that they were about the size of fox terriers, emerged from the den and immediately began to suckle. We guessed the cubs were about eight weeks old. After 15 minutes the mother abruptly terminated the suckling by rolling onto her stomach. Then she stood up and dug at the hole for a short while, before moving off. As she left the cubs went back inside.

For the next three months we made regular observations at the den and tracked the spoor of D'Urbyl, as we named the mother. The name was a combination of the names of Margie's mother (D'Urban) and mine (Sybil). Unlike the mother in the study at Cubitje Quap den, the mother of these small cubs made regular visits to her den, at sunrise and sunset, to suckle her cubs. Her visits were short, usually not lasting more than half an hour. The cubs rarely came out of the den when their mother was absent. The den itself was quite small, too small for the adult to enter, but it obviously provided good protection for the cubs for the long periods that she was away. All hyena mothers are excellent when it comes to nursing their cubs. They have an extensive lactation period of over a year and their milk is of a very high nutritional standard. Suckling bouts may last up to an hour with large cubs, when all parties concerned seem to fall asleep in what appears to be a blissful state of domestic bonding between mother and offspring. I am pleased to say that my wife took their example to heart with our own children.

When the cubs were about four months old, a number of things began to change. Firstly, D'Urbyl now usually made only one visit to the den every 24-hour period, usually at night. Secondly, we started noticing the remains of food items at the den, obviously brought there, which we had not seen before. The third change was that the cubs started to spend periods out of the den when their mother was not there. Typically, they would emerge at sunset and spend an hour or so sniffing around, picking up scraps of food and playing with each other. They also became used to the Land Rover and we were soon able to approach close enough to discern that they were two males and a female. Our excitement at being able to watch them at close quarters helped to heal our wounds from the previous disasters.

Early in January 1973 a significant event took place in the study. We were visited by Dr Hans Kruuk. Hans was the man who during the 1960s had revolutionised the understanding of hyenas through his classic studies of the spotted hyena in the Serengeti region of Tanzania, and later through his studies of the striped hyena and the aardwolf. Hans was to become my mentor and friend and was to play an important role in my hyena studies from then on.

My biggest problem at this time was to catch hyenas by an acceptable method. Hans said that he had had success in catching striped hyenas in cage-traps. He saw no reason why brown hyenas should not also be tempted to enter them. We managed to get hold of an old cage and fix up a rather crude drop-door for it. Next we had to find some bait. As fate would have it, we found a half-eaten duiker carcass in a tree. In the interests of science we stole the carcass from the leopard. The idea was that the hyena would enter the trap, pull the meat, which was attached to a peg holding up the drop-door, and the door would drop closed. We set the trap near Cubitje Quap Windmill.

After supper that night we decided to pop out to see if anything was in the trap. As we approached we were excited to find the trap-door closed. The excitement soon changed to apprehension when we were greeted by an ominous, deep-throated growl. This did not sound like a hyena, nor was it one. It was a very large and unimpressed male leopard – poetic justice for having stolen the carcass from a leopard in the first place!

How were we to get the leopard out of the trap? Hans suggested that I drive the Land Rover up to the trap so that the back wheel was against the door. He would climb on the back of the truck, pull out the gate and I would drive off as fast as I could. It seemed like a pretty risky option to me, but who was I to argue with a man of his experience.

There are few noises in nature more intimidating than a leopard's growl. I drove the Land Rover into place, and waited for Hans to get ready, my foot on the clutch shaking involuntarily with fear. Hans pulled the gate and I accelerated. Then I heard him shout: 'No, no, no!' Oh hell, I thought, the leopard has jumped onto the back of the truck and got hold of Hans. In the event I had pulled off too quickly and Hans had not been able to pull the gate out completely, so we had to go through the whole procedure again. This time everything went according to plan and the leopard bolted off.

The next day we rebaited the trap with some springbok. After four days we had still not caught anything and were beginning to wonder if indeed brown hyenas were trappable. On the fifth morning the trap-door was down. We approached on foot, as we did not want the animal, if it was a brown hyena, to associate being trapped with the vehicle. Our hopes were realised when from a distance we made out the shape of a brown hyena lying quietly in the trap. Apart from a very brief period when I walked up to the trap to fire the dart at the animal from the Cap-Chur pistol, the capture procedure was quite untraumatic. We attached a radio collar to the hyena, a young female whom we called Normali, after my sister-in-law Norma and Margie's sister Alison. We also cut a small nick out of her ear for permanent recognition. Once we had finished working with her we placed her back in the trap, which we then covered with grass, and left her to recover from the drug in peace and safety. When we released her the next evening, she ran off completely recovered. Our euphoria at being able to catch hyenas safely was soon tempered by the fact that we were unable to relocate Normali. It appeared as if the radio had also failed within days of being fitted. Little did we know, however, that Normali was to become a very important character in our study in the future.

Cage trapping was obviously a far more satisfactory capture method than the one we had previously used. Not only was the actual capture relatively untraumatic, but we were also able to keep the animals safe until they had fully recovered from the immobilising drug. A cousin of Margie's, Jack Rivett-Carnac, kindly made four traps for me at his engineering works in Johannesburg. Over the next five years I was to catch 29 different brown hyenas in these traps. I also caught another four leopards, six lions and numerous jackals. While the jackals could escape through the bars, the larger carnivores were released from the safety of the vehicle by means of a rope attached to the trap-door and passed through a pulley in the tree above the trap.

The only drawback to the cage trap was that it was non-selective. Some hyenas were trap-shy and never entered it. Others were trap addicts. The biggest addict was a young male called Bing whom we caught 17 times in a year! He was named after a friend Bing Lucas, Director of New Zealand Parks, who was staying with us when we caught him for the first time. The capture procedure was also soon made easier by the availability of two new drugs called ketamine and zoletol, which kept the animals down for far shorter periods than Sernylan.

For about a week around each full moon we held all-night vigils at the Kaspersdraai den. We had also managed to trap D'Urbyl and fit a radio collar on to her. This one worked for several months, but the range was not good and we were rarely able to find her except when she visited the den. However, we did log up over 400 kilometres of tracking her spoor and were able to make the first measurement of the size of a brown hyena's territory. As with all brown hyenas we had tracked, she always moved on her own.

The cubs soon became totally habituated to our presence and we were able to observe them as close as we liked. The highlights of these nocturnal vigils were D'Urbyl's visits to the den. When her cubs were four to nine months old she visited the den to nurse them on average three nights out of four. She was not nearly as relaxed in our presence as her cubs were, so we had to stay back when she was around. Occasionally she would arrive with some food: we saw her bring bat-eared fox, black-backed jackal and steenbok carcasses, and she always suckled the cubs for about half an hour. However, in contrast with our previous experience at the Cubitje Quap den we never saw another adult visit this den.

Much of the time was spent waiting for things to happen. Margie whiled away the time by crocheting in the moonlight during her watches. I had no such aid for keeping awake and was far more likely to fall asleep on the job than she was.

When the cubs were about nine months old they began to wander a bit from the den. Typically they would move off at sunrise for about a kilometre before settling down for the day under a bush. Then in the evening they would return to the den by a slightly longer route. We began to follow the cubs in the Land Rover during these first foraging expeditions. As they grew older they began to increase the distance they moved, quite often moving 10 or more kilometres in the day. Like the Cubitje Quap cubs, these always moved on their own.

Hans Kruuk had suggested that I should try to follow hyenas at night. If this could be done, it would open up a whole new range of possibilities for the study. Having, as it were, served our apprenticeship with the cubs in the daytime, I was anxious to see how we would fare at night. We began to gain experience in following them after dark on some of their short return journeys to the den from a day-time resting site. If we lost the hyena we returned to the den and waited there. We used a small hand-held spotlight to illuminate the hyena and the vehicle's parking lights to see our way. The main problem with using lights to follow carnivores is that they may dazzle the prey and make it easy for the predators to make a kill. This is obviously undesirable as a researcher wants them to behave as they would if no human was present.

Eventually we decided to try to follow a cub when it left the den after dark. We chose a bright moonlit night, as we thought that this would make things easier. At 8.10 pm the female cub, whom we had called Bop, moved off. With our eyes glued to the little hyena we started to follow her in the Land Rover. In the moonlight we were able to follow what looked like a shadow gliding through the sparse vegetation. Whenever we lost visual contact I quickly switched on the spotlight to find her again. On and on she moved, occasionally stopping either to sniff upwind, with her head held high, or to smell something on the ground, with her nose down.

One hour slipped by, then another. According to the vehicle's odometer we had travelled five kilometres. Suddenly Bop darted to the left. I swung the spotlight round and we saw her, 10 metres away, biting at something small. We drove closer to see that she had caught a striped polecat as it dashed for a small hole. She was biting it in the head as the animal wriggled from side to side. Soon it stopped moving and Bop let go her grip. She put the animal down and started eating it head first. Thirty minutes from the time she killed the polecat, all that was left in Bop's mouth was its tail. With two bites she swallowed this and with a flick of her own tail moved off. If we had tracked Bop's spoor the next day, all that we would have found here would have been a small patch of blood and a few hairs. There was no way that we would have been able to reconstruct these events.

As we were to find out, striped polecats are very unusual food items for brown hyenas. Their remains were only found in 0.3 per cent of the scats we analysed. On only two other occasions during the eight-year study period did we see a brown hyena hunting these little carnivores of the mustelid family. Once a brown hyena stuck its head into a small bush out of which ran a striped polecat, but the hyena let it go. On another occasion a brown hyena lunged rather half-heatedly at a polecat, which turned to face the hyena, repeatedly approaching and backing away from it with its tail curled forwards over its body in a threat display. In both these situations the brown hyena seemed to be in a good position to catch the polecat but, unlike Bop, failed to do so. Perhaps striped polecats are not very appetising for brown hyenas and are only killed when other food is scarce. As we were to find out a few days later, the cubs' mother was dead. Perhaps without her mother to help provide food, Bop was in a more desperate position than we realised.

Bop continued walking until 3 am. During this time she found two patches of protein-rich harvester termites and spent half an hour at each, licking termites off the sand. Then she lay down. We were relieved to have a break and enjoyed a most welcome cup of coffee while we waited to see what she would do next. However, little Bop was finished for the night and as the sun began to rise we realised that, perhaps for the first time in her life, the cub was not going to return to her den that night. Before setting out with Bop, we had planned to return to the den and spend the day at the makeshift camp we had erected there. We had no idea how to get back to the den from where we were, so decided to sleep the day there and to follow our cub the next night. Although we didn't have a very comfortable day trying to sleep without proper camping gear, we were well contented with the night's activities.

Soon after sunset Bop was on the move again and we continued our journey. Just before 8 pm the inevitable happened – we lost her. She had moved up a fairly high dune, but as we came over it there was no sign of the hyena – she had simply vanished into the night. After searching with both our hand-held spotlights for over half an hour, we reluctantly had to admit defeat. All we could do was to take a westerly bearing and drive until we came into the Nossob river-bed, which was probably some 30 kilometres away. We set off in that direction feeling pretty down-hearted. What an anti-climax after the previous night's success! However, we hadn't gone a kilometre when we saw some eyes shining in the lights ahead of us. Miraculously it was Bop, plodding along in front. She didn't even look at us as we fell in behind her again. Soon after midnight she brought us back to the den.

We began to build up a record of what the cubs were eating. As the spoor tracking data had already suggested, scavenged vertebrate remains and tsama melons were important food items, but insects, such as the harvester termites we had seen Bop eating and large scarabid beetles, were also important. Moreover, we were able to measure the amount of food they obtained far more accurately than we could when following spoor. Bop's polecat kill was obviously exceptional and for the most part the cubs showed little interest in hunting.

One morning a few days after our adventure with Bop, we followed one of the male cubs from the den. Originally we called this cub Nick because of a small natural nick in his ear. Instead of the usual unhurried, rather meandering manner in which the cubs foraged, Nick seemed to be more determined and kept going in one direction as if he knew where he was going. After he had moved nearly five kilometres in this manner, he stopped and raised the long hair on his back – a sure sign of something unusual or frightening. He moved forward slowly, his mane still bristling. We could smell something really rotten. Then we saw what the cub had sensed – a dead brown hyena. The cub went up to the carcass and started pulling at it. The hindquarters had already been eaten and he started eating at the ribs. It was then that we noticed the radio collar – he was eating his mother! D'Urbyl had obviously been dead for several days and we were unable to find any clues as to the cause of death. At first we were quite shocked at the cub's behaviour, but when we analysed it rationally it really made a lot of sense. If you are a brown hyena cub and your mother dies, the best thing you can do is eat her, although I doubt very much that the cub actually identified the carcass as her mother's.

The important question now was: would the cubs be able to survive without their mother, especially as there were no other brown hyenas in the territory to help feed them? In fact the cubs managed quite well for the next six weeks when they abandoned the den, slightly earlier than normal. After this we lost contact with them. However, a week later we found the remains of a brown hyena cub, which had been killed and partially eaten by spotted hyenas at Kaspersdraai Windmill. It may have been one of the cubs. Later we were to learn much more about the relationship between the two hyena species.

Soon after this we found our third den. This one was closer to home, about five kilometres in the dunes from Rooikop Windmill, just south of Nossob Camp. It wasn't long before we managed to trap the mother of the two cubs, an old female whom we called Ro-Ro, after Margie's brother Rob and his wife Roz. At this time I had no more radio collars, so we merely clipped a small section of her ear so that we could identify her in future. I had previously fitted my last radio collar to a male, whom I trapped at Nossob Camp. This male was called Jo-Ro after another of Margie's brothers and his wife, John and Roz. Jo-Ro was subsequently seen to bring food to the Rooikop den.

Although we did not spend as long at this den as we had at the Kaspersdraai one, we did make one very significant observation here. We once saw a third adult visit the den, which, judging by the reaction of the cubs, was known to them. After 18 months of study we had never seen two adult brown hyenas together, nor had we tracked the spoor of two moving together. Yet we had evidence that brown hyenas co-operate in raising their young, as when the males brought food to the Cubitje Quap and Rooikop dens, and now we had even seen a third adult at a den that seemed to be known to the cubs. If brown hyenas were monogamous, what had happened to D'Urbyl's mate, and who was this third character? It was to be several years before we began to find answers to these puzzles.

In the meantime, flushed with our success in following Bop, we decided to try to follow the male Jo-Ro at night. The radio collar would, hopefully, enable us to relocate him whenever we lost visual contact. One bright moonlit night when we were at the den he visited it and presented us with the opportunity we had been waiting for.

Following Jo-Ro, we quickly discovered, was completely different from following Bop. In the first place the adult moved faster than the cub, but, more importantly, Jo-Ro was not nearly as well habituated to the vehicle and ran off if we approached too closely. Within minutes of his moving off from the den we lost visual contact. Margie jumped on to the back of the Land Rover, unfolded the large and cumbersome Yagi antenna, and held it aloft so that I could ascertain the direction in which to drive from the radio signal. Before I drove off she had to fold up the aerial, as she was unable to keep it up and hold on at the same time with the vehicle bumping over the dunes. For the next three hours we struggled to regain visual contact with our quarry. I simply was not able to hold the correct direction for long enough for us to find him, before he changed direction slightly or ran away from us. It was only at sunrise that we eventually saw him again. Now he was even shyer of the vehicle than when it was dark. We could only follow him at a distance and had to be content to watch him from the top of one dune to the other. Eventually he disappeared under a tree where we knew he would rest for the day.

The next evening at sunset we returned to continue the unequal struggle. We hoped that we could habituate him to the vehicle. Even if we were unable to maintain visual contact all the time, at least we might be able to obtain a good idea of where he moved. Unfortunately things were no better this night. As soon as it became dark we lost him. As we couldn't relocate him, we were unable to habituate him. We seemed to be caught in a vicious circle. With Margie gamely operating the Yagi antenna from the back of the Land Rover, we were able to establish that he was moving towards the Nossob river-bed. Perhaps we would have more success in seeing him in the more open and flat habitat.

As he seemed to be moving towards Kaspersdraai, we decided to go to the windmill and wait for him there. On driving up to the waterhole, we sighted five pairs of shining eyes in the vehicle's lights. The eyes belonged to spotted hyenas. We keenly waited to see what would happen when our brown hyena met up with the spotteds. The signal from Jo-Ro's radio indicated that as he approached the windmill, he circled round so that he was down wind from it. Then he obviously sensed that the spotted hyenas were there, as he apparently backed away. For the next 35 minutes the signal stayed constant while the spotted hyenas lay close to the windmill, oblivious to the close proximity of the brown hyena. Then the radio signal told us that Jo-Ro was moving away. His need to drink was obviously less urgent than his wish to avoid an encounter with the spotteds.

By the time we realised that he had moved off, Jo-Ro was some distance away. Once again we attempted to locate him, but all we could do was to maintain radio contact, even though he was obviously moving along the river-bed. We were having no more success here than we had had in the dunes. He moved all the way to the next windmill at Rooikop, 18 kilometres away. As we drove up to the windmill, with the radio signal getting louder and louder, we were surprised to see two pairs of eyes coming from the direction of the radio signal. There was Jo-Ro standing eyeball to eyeball with another brown hyena.

Neither animal seemed disturbed by the spotlight, so intent were they on sizing each other up. Suddenly the stranger lunged at Jo-Ro, grabbed him by the side of the neck and started dragging him around. Jo-Ro, his ears back and hair raised, protested loudly at this treatment, giving vent to a deep snarl and sharp yell that reverberated into the night. The harder the stranger tugged, the louder Jo-Ro yelled. After half a minute he managed to break free. The stranger, however, stood over him with his ears cocked and looking more aggressive than any brown hyena we had ever seen. With his mouth wide open, ears still flat against his head, Jo-Ro peered up at his dominant rival. The two then proceeded to pitch their heads from side to side and parry and thrust at each other's mouths, without making contact, though Jo-Ro continued to snarl loudly.

After a minute of this they stopped. Jo-Ro stood up and moved away, bleeding slightly from the side of the neck, his tail tucked under his belly. The other watched him go, standing erect and pawing the ground. After moving about five metres, Jo-Ro started running, and his rival moved off slowly in the opposite direction.

If we had had difficulty in keeping up with Jo-Ro previously, it now became impossible. He was obviously making a bee-line for the centre of his territory. We managed to follow his radio signal for another hour, but eventually, from a combination of high dunes and fatigue, we lost radio contact.

It was a great disappointment and most frustrating. The previous two nights had been exhausting and we had not really succeeded in our objective. I felt sure that it was possible to improve. The fight with the unknown hyena had certainly been the highlight, but it left many unanswered questions. Who was the strange hyena and why had they fought? Was this a territorial dispute? If so, why had the victor let his subjugated rival go so easily?

What we needed was more practice in following hyenas and, even more importantly, a better radio-tracking system. The transmitters were too unreliable and their range of about one kilometre under normal operating conditions was inadequate. Furthermore, the large and unwieldy Yagi antenna was no good for our needs. We needed a directional antenna system that was easier and lighter to use, preferably one that I could stick out of the Land Rover window while listening to the signal through earphones and driving the vehicle.

We arrived back at Nossob Camp at daybreak and tried to get some sleep before the heat in the caravan became too intense to bear. That day the wind started blowing strongly from the north and by mid-day it had developed into a severe dust storm. Not only did the caravan and tents become covered in layers of dust, but I began to worry that the canvas might tear. Later in the afternoon thick clouds began to form and by five o'clock the heavens opened in the most spectacular thunderstorm I had ever experienced. With the wind lashing and the rain coming down in buckets, the flimsy canvas eventually yielded to the forces of nature. The bottom of our large living tent tore. To make matters worse, a large pool of water formed in the hollow outside our camp, which then poured in through the torn hole. We frantically tried to rescue as many of our belongings as possible, but there was nothing we could do to stop the tent from tearing further. After a little over an hour, during which 38 millimetres of rain fell, the storm passed and the wind dropped from a roaring gale to a stiff breeze. Elias and Doempie le Riche kindly took us into their house for the next few days while we sorted out the mess and set up a new camp minus one of our tents.

It was now early November and 18 months of my two-year study had elapsed. In the meantime the National Parks Board had advertised a new position of research officer for the Kalahari Gemsbok National Park. I was very interested in this post. Not only would it mean that I could expand my brown hyena study, but Margie and I had grown to love the Kalahari and were keen to extend our time there. As is so often the case with a lucky break, I happened to be in the right place at the right time and was thrilled to hear in January that I had got the job. At that time my appointment raised some eyebrows – a liberal 'Engelsman' employed in an Afrikaner kingdom. Perhaps it was a case of rather the devil you know than the devil you don't. I was to take up the position in June 1974. One of the conditions was that Margie would be required to help out by taking over the running of a camp from time to time when the camp manager went on leave.

Now that I was certain to be able to extend the brown hyena study, I set about trying to obtain funding for a new and more efficient radio-tracking system. The Endangered Wildlife Trust (EWT) agreed to support this aspect of the study. This was the beginning of a long and, for me, most fruitful relationship with the organisation. Subsequently they were to make very significant contributions to my studies on cheetahs and wild dogs in the Kruger National Park. This culminated in an even closer relationship between 1998 and 2005, whereby, through an agreement between SANParks and the EWT I was able to carry out my dual role as specialist scientist with SANParks and head of EWT's Carnivore Conservation Group.

Almost a year was to elapse before I finally received the new radio-tracking system. During this time we analysed in detail the data we had accumulated until then and I settled into my new job. We also spent a six-week period searching, in vain, for evidence of brown hyenas in the Kruger National Park and, as I will relate in the next chapter, had a short interlude getting to know spotted hyenas in the Auob river-bed.

CHAPTER 3

Some Spotted Snippets and a Brush with Rabies

The storm that wrecked our tent heralded the beginning of an exceptionally wet summer for the southern Kalahari. Over 600 millimetres of rain fell at Nossob Camp between October 1973 and April 1974, compared with the annual mean of 220 millimetres. As is always the case in arid regions, the vegetation responded to this bonanza in a most spectacular way. The desert bloomed. Flowers and annual grasses we had not seen before sprang up all over the place. Large stands of the annual Kalahari sour grass appeared like fields of corn. Yellow *duiweltjies* and cleome, purple catstail, and pink and white *driedorings* added colour to the vista of green. Kwang Pan, which had always been brown and dry, was covered in a carpet of grass, on to which poured a thousand or more springbok, providing a spectacle to rival any natural panorama. Birds such as marsh owls and monotonous larks, whose distribution maps in the bird books gave no hint of their occurrence in the southern Kalahari, became abundant for several weeks, before disappearing once again.

I decided not to continue with tracking brown hyena spoor as I felt that we had extracted as much information from this technique as we could. In any case Houtop, after doing a sterling job of following nearly 1 200 kilometres of spoor, had decided he had had enough, and who could blame him?

Now that I was employed by the National Parks Board, my research interests in the Kalahari became wider than just brown hyenas. I set up a monitoring system of the antelope in the area by doing monthly ground counts along the river-beds and twice-annual aerial surveys over the entire area. I also started looking at other carnivores, particularly spotted hyenas.

Spotted hyenas are rare in the southern Kalahari: there are twice as many brown hyenas. Both Elias and his brother Stoffel, the park warden, who had lived all their lives in the Kalahari, believed that spotted hyena numbers had declined in recent years. In spite of their low numbers, spotted hyenas are often easier to locate than are brown hyenas, because they regularly rest along the river-beds, which brown hyenas hardly ever do. I was asked to do a preliminary study of spotted hyenas, in order to get some idea of their numbers. Later I hoped to initiate a more detailed study of these still largely misunderstood carnivores. Apart from the question what factors were limiting their numbers, there was the still controversial issue of how much hunting they do. Evidence gleaned from following spoor in the Kalahari suggested that spotted hyenas there were indeed active predators capable of pulling down such formidable prey as adult gemsbok.

I soon found that there were at least four spotted hyena clans living along the Nossob river-bed and two in the Auob. We found a den near Kaspersdraai – the hyenas that had kept Jo-Ro from the windmill and that had killed the cub we thought was Bop. There were at least three breeding females with cubs at this den, as well as several other animals in the clan.

We found a second den near Urikaruus in the Auob river-bed. Two females, each with one small cub, a sub-adult and an adult male, made up this small clan. As we had to spend a month at Mata Mata Camp relieving the ranger Isak Meyer and his wife Annie, who were due to go on leave, I decided to use this time to get to know the Auob system and these hyenas better. Margie, however, was often confined to the camp as acting camp manager.

I wondered how easy it would be to follow spotted hyenas at night. If they moved along the river-bed and in a group, I reasoned, it should be easier than it was to follow the shyer and solitary brown hyenas through the dunes. This proved to be true up to a point, and I spent several moonlit nights with the Urikaruus four. However, they didn't stick only to the river-beds, and the dunes in this part are high and very difficult to negotiate. Furthermore, they moved erratically, sometimes running off at speed, and when this happened I invariably lost them.

On one occasion, early in the morning, they ran up to a gemsbok cow, which turned around and fled. At this the hyenas, strung out in a long line, gave chase. Forgetting for the moment about holes and hidden logs, I put my foot down and went after them, adrenaline pumping in the excitement. About one kilometre on, the gemsbok backed up against a shepherd's tree with the hyenas making a determined effort to get at it. Each time a hyena lunged at the gemsbok, she lowered her head and the hyena jumped back, obviously to escape the lethal horns. After a minute or so the hyenas gave up and walked away slowly. Was this a rare attempt to catch such large and formidable prey or was it, as the evidence from tracking spoor suggested, normal behaviour for Kalahari hyenas? It was to be several years before I was to discover the true facts, as our month was nearly up and it was time to return to Nossob and the brown hyenas again.

Before this happened I made one other very significant observation. I arrived at the Urikaruus den late one afternoon to find one of the adult females and the sub-adult lying there, with fresh blood around their faces. I thought that they must have just made a kill. I searched around for evidence of a carcass, but found nothing. Besides, their stomachs were not full, which I would have expected if they had just had a good meal.

Suddenly the sub-adult jumped up and attacked the adult, biting at her hindquarters. The adult did not retaliate but, crouching to protect herself, merely moved away a short distance. Then the sub-adult started lowing – a common spotted hyena vocalisation, not unlike the lowing of a cow, but deeper. This seemed to upset the adult, for she ran over to the youngster, though not aggressively. At this it attacked her again and again the female merely protected herself without retaliating.

At this moment the second adult female from the clan appeared. The two adults engaged in the strange spotted hyena meeting ceremony in which they stand head to tail and sniff at each other's sexual organs, and one of them whooped several times. (The whoop is the characteristic spotted hyena call about which I will have more to say later on.) The sub-adult ignored the calls but, whooping itself, ran over to the den entrance from which one of the small cubs was emerging. At the sight of the demented hyena bearing down on it, the cub retreated into the den. At this the sub-adult spun around and attacked the females. It bit at their hindquarters as the females took evasive action by crouching and spinning round to face their attacker. Then the sub-adult bit at their necks and mouths. They counter-attacked, one of them grabbing the unfortunate youngster by the side of the mouth and shaking it vigorously for several seconds.

Next, the adult male of the clan was seen approaching the den. Immediately the sub-adult rushed towards him and the adult turned tail and ran away. The sub-adult continued running, even though it had obviously lost interest in chasing the male. It continued at a lope for two kilometres until it came to Urikaruus Windmill. Here it went right into the drinking trough and attempted to drink, but it was unable to lap the water and only managed to scoop a small amount into its mouth. For the first time I was able to have a good look at the animal. It had a dazed and disorientated expression, seeming not to see me even though it looked at me. As it walked off it stumbled. I followed it for a short while before leaving it.

I have no doubt that the sub-adult had rabies. Its aggressive behaviour towards animals that were dominant to it, its disorientated state and its inability to drink were all classic signs of this dreaded disease. I never saw this animal again, even though I spent several nights at the den during the next two weeks.

Even though I was not planning to make detailed observations of these hyenas once we returned to Nossob, I had decided to try to catch and ear-clip some of the animals for future recognition. Accordingly, I had brought one of my traps from Nossob and set it close to the den. I was surprised that nothing went into the trap, in spite of the tempting piece of fresh springbok haunch I had placed in it.

However, three nights after the drama at the den I caught the adult male. This, incidentally, was the only spotted hyena I ever caught in a cage trap. I duly processed him in the usual way: took body measurements, examined his teeth for ageing purposes and cut a small notch in his ear. Shortly afterwards it dawned on me that this animal could have contracted rabies. Although I never saw him being bitten by the suspected rabid animal, it was not impossible that at some time he had been. Rabies can only be transmitted through saliva entering the bloodstream, usually from a bite. When I arrived home after handling the animal, I noticed that I had some small open sores on my hand, inflicted from transporting and setting up the trap. I began thinking: If the animal did have rabies, and if I had got some saliva on my hands when examining his teeth, could the virus have entered my bloodstream? It all sounded a bit far-fetched, but on the other hand I didn't want to take any chances.

To set my mind at rest the next morning I drove off from Mata Mata to see the district surgeon in Upington, 400 kilometres away. I was told that the incubation period for rabies was highly variable and that the symptoms only became apparent once the virus reaches the brain. The speed at which this happens depends mainly on where and how badly the victim is bitten by the rabid animal. If, as with the hyenas at Urikaruus den, it is bitten around the mouth, it can take a few days for the virus to reach the brain and for the symptoms of the disease to be displayed. It is only when the symptoms are being displayed that rabies can be transmitted by an animal, and this does not usually last more than about 10 days before the rabid animal dies. Although the district surgeon admitted that the chances of my picking up the virus were small, if the animal did die in the next 10 days I should have to undergo the notoriously uncomfortable and traumatic anti-rabies vaccination course of 14 daily injections as a precaution.

So back to the Kalahari I went to keep an eye on the hyena. The problem was that the male was not as regular a visitor to the den as the two females were. Anyway he was at the den on the fourth, fifth and sixth nights after I had handled him, and neither he nor any of the others showed any signs of unusual behaviour. I did not visit the den on the seventh and eighth nights but went again at sunset on the ninth night. The two females were there, but when I left several hours later the male had not appeared. On the tenth night at sunset, again only the two females were at the den. Shortly after dark they moved off. I waited two hours, but nothing came. I wasn't really worried, but I would have been very pleased to see my friend. I knew that the chances were good that the hyenas wouldn't return to the den until the early hours of the morning, so I decided to pack it in and go home. As I drove off I caught the reflection of some eyes in my lights. It was a hyena. I picked up my binoculars and was very relieved to see the nick I had cut in the right ear 10 days previously.

Two days later we returned to Nossob Camp. It was to be another two weeks before I could visit the Urikaruus den again. When I eventually did, it was deserted. In fact, I never saw the three remaining adults and the two cubs again. Although I will never know for sure, I can only assume that all of them succumbed to rabies. Why else would they disappear from an area in which they were obviously so well established? Even more significantly, it was to be another 10 years before spotted hyenas were again seen regularly in the northern reaches of the Auob and 35 years before spotted hyenas were again seen using the den.

At about this time too, spotted hyenas disappeared from the vicinity of Nossob Camp, where, although it was never very common, we had regularly heard them calling, and on occasions even seen one or two as well. Here too it was to be nearly 10 years before they returned to the vicinity of the camp on a regular basis. I have no evidence that rabies was involved here, but even if it was not, it is significant that it took so long before spotted hyenas returned. But more about that later. By the end of the year, my new radio tracking equipment had finally arrived. At long last we were able to continue with the brown hyena study.

CHAPTER 4

Progress
and Setbacks

HANS KRUUK had written to me early in 1974 about beta lights. The manufacturers of beta lights advertise them as 'lights without power'. They are sealed glass capsules coated with phosphor paint and filled with tritium gas. Tritium is an isotope of hydrogen and, therefore, mildly radioactive. The hydrogen electrons explode against the phosphor paint, causing the beta light to glow like a bunch of glow-worms. Tritium has a half-life of 15 years, which means that after 15 years a beta light will be half as bright as it was when manufactured. Hans had attached beta lights onto radio collars he had fitted to European badgers and had found them a great help in maintaining visual contact with his study animals. He suggested that I might try them on brown hyenas.

I suspect that the South African agent thought I was trying to be funny when I told him that I wanted to put beta lights on brown hyenas. Once I had convinced him that I was serious, he informed me that before he could get them for me, I would have to obtain permission to use them from the Atomic Energy Board. After a fairly lengthy correspondence, I was finally given permission to use the lights, provided I attached a notice to each collar saying: 'DANGER RADIOACTIVE: If found return to Warden, Kalahari Gemsbok National Park'. Apparently there is more radioactivity in the luminous paint used in a watch than there is in a beta light.

Armed with five new radio collars, a receiver, a directional loop antenna, which was far lighter and unwieldy than the old Yagi, and 10 beta lights, together with the required warning notice, we were ready to re-start the brown hyena study in early December 1974. Within days we caught Normali at Cubitje Quap. Normali was the first brown hyena we ever caught in a trap, but we had not seen her since the day we released her with one of the old radio collars, nearly two years previously. In the meantime she had shed her radio collar. We were excited to see that she was lactating.

After fitting the new radio collar, we placed her back in the trap to allow her to recover fully from the drug. We opened the cage the next morning but she didn't realise that it was open, and when we went back in the evening she was still there. I moved the grass that we had put on and around the cage to give her good shade, and only then did she realise that she was free. She ran out of the cage and over a small dune into the river-bed. By the time I climbed back into the vehicle and drove to the crest of the dune, she had gone. Here was our first chance to test the new radio-tracking system. The loop antenna was mounted on a piece of hardboard behind the cab of my new vehicle, a Toyota Land Cruiser. Margie hopped onto the back and handed it to me. In turning the antenna, the signal strength changed. When the antenna was pointing at the animal the signal was strongest; when the loop was at right angles, the signal was weakest. The only trouble was that there was no front or back. When the signal was strongest the hyena was either in front of you or behind you. This time it was simple as we knew in which direction she was to be found. We were relieved to quickly find her moving in the fast-fading light over to the windmill where she had a long drink. We waited at a good distance away, not wanting to disturb her. By the time she had drunk her full it was almost dark. We were thrilled to see how clearly the beta lights shone in the darkness and were able to follow them as she moved along the flat and open Nossob river-bed, with only the car's parking lights on to detect any obstacles. She seemed unconcerned by our presence and during the next hour I gradually narrowed the distance between her and the vehicle. This was most encouraging.

We expected that Normali would return immediately to her den.

She had been away for at least a night and a half, which, although not excessive, was nevertheless long enough if her cubs were small. As it turned out, she was obviously in no hurry to return to her cubs and took us on quite an extensive journey up the Nossob river-bed, across Kwang Pan and into the dunes on the Botswana side. This was useful as it served to habituate her to the vehicle and gave us valuable practice in following her. With one or two minor hiccups, mainly when we encroached too close for Normali's peace of mind and she ran off a short distance, things went well. She chewed on a small bone for two minutes and then ate something small which we were unable to identify. She then came across a beehive in the base of a shepherd's tree and proceeded to devour the honeycomb until the bee stings became too much for her to handle. Further on she found a fairly old carcass of a gemsbok calf and chewed on the dry skin and bones for over half an hour. Finally at 3.30 in the morning, after we had followed her for 23 kilometres, she came to her den.

Her arrival was most abrupt. Suddenly, there she was on the mound. Luckily we were not too close at this point and she was unconcerned. Within a few seconds three cubs of about six months old emerged, and she immediately moved off in the long grass with them. However, from the radio signal we knew that she had lain down close by and was suckling the cubs. After 27 minutes a cub appeared on the mound, followed by the other two. We could also see Normali's beta lights shining in the grass behind the den. The cubs were not too sure about us but stayed above ground, sniffing around on the mound and in the grass. Normali moved off just before sunrise, but we stayed behind. I wanted to establish where we were in the daylight, so that we could make a track from the Nossob River to the den. Not being proficient at navigating at night through the dunes, and not knowing this area very well, we weren't at all sure how far from the river-bed we were.

By the time the sun rose the cubs had all returned to their den. We drove off in a westerly direction for the Nossob river-bed. I thought that we must be all of 10 kilometres from the river, so I was pleasantly surprised when, after only 2.6 kilometres, we came over a dune and there was the river-bed. We were roughly half-way between Cubitje

Quap and Kwang Pan, less than half an hour's drive from home.

We climbed into bed that morning feeling really contented. The radio-tracking system was a vast improvement on the old one and the beta lights were fantastic. Hans Kruuk had said to me that we were sitting on a gold mine of information about the brown hyena, and for the first time I felt that the gold mine was moving into operation.

Over the next few weeks we spent many nights following Normali. As I suspected, she spent a lot of the time foraging along the river valley, although she also used the dunes extensively. By using the stars and moon as navigational aids and recording the distance from the Land Cruiser's odometer, we were able to plot these movements on a map as we had recorded the movements of the other hyenas from tracking spoor. Twice she met up briefly with another hyena that limped badly on his left back leg. In February we caught this animal, an old male, whom we called Hop-a-Long and fitted a radio collar to him as well. Hop-a-Long was a wily old fellow whom I only managed to trap twice in four years.

We also retrapped and collared the old female from the Rooikop den, Ro-Ro, and another two females that lived in an area between Normali's and Ro-Ro's territories, in much the same area as the fateful Cubitje Quap animals. One of these was an old female, whom we called Cicely, after Margie's grandmother. She had a single large cub in a den about seven kilometres north-west of the old Cubitje Quap den towards Seven Pans. The second female was younger, called Floppy-Ears on account of one tatty ear which could not stand upright. Although it appeared that she had had cubs, Floppy-Ears did not produce cubs in the 12 months that we tracked her. We also never saw Cicely or Floppy-Ears associate with an adult male, nor did one ever visit the den while we were there.

Floppy-Ears was one of the trap addicts. I caught her 11 times in eight months. Once I even followed her from one trap to another and watched her go into the second trap. At first she tried to get at the meat from the back of the trap but gave up fairly quickly, experience obviously having taught her that this was impossible. She then stood at the entrance for some time pawing the ground in frustration.

Eventually she could stand the temptation no more and went in to claim her free meal.

Floppy-Ears was also the most laid-back brown hyena I ever encountered, even when I approached her on foot in a trap. When I darted her she did not appear to have received the full drug dosage. She lay quietly in the trap, apparently only half-drugged, but sufficiently drowsy to allow my father-in-law to sit down next to the trap and tickle her on the nose with a piece of grass. Eventually I decided to give her a bit more of the drug and administered it with a syringe by hand. When she eventually went under and we were able to pull her out of the trap, we were amazed to learn that the original dart had not gone off. We had tickled her nose and hand-injected her while she was completely awake.

The first half of 1975 was a most productive period. However, it wasn't long before Murphy's law struck again. In June the radio receiver packed up, so we were unable to locate any of the radio-collared animals on a regular basis. I had to send it back to the US to be repaired and it was only returned to me in December, this time with a back-up receiver as well. Frustratingly we had to watch the year go by and were only able to follow a hyena opportunistically, if one came to a den we were at, or, as with Floppy-Ears in particular, when we managed to trap one. Even then we only had the beta lights to help us maintain contact. If we lost visual contact with the hyena we were following, we were unlikely to relocate it that night.

Nevertheless it was during this time that I made one of the most exciting brown hyena observations of the entire study. Unfortunately Margie was not with me that night as she was helping out at Nossob Camp. I had caught Cicely in a trap out in the dunes near Seven Pans and followed her on release. Soon after dark she began to meander with her nose to the ground. For half an hour she moved slowly, often circling back to the same place, as if she was following a scent trail. Suddenly all the hair on her back bristled as she came upon an ostrich nest with 26 eggs in it.

After sniffing at the eggs for a few minutes, as if she could not believe her luck and was counting them to make sure, she picked one up and carried it off 50 metres before dropping it in an open patch. She then ran back to the nest where she ate two eggs, easily opening them by biting a small hole at the point and lapping up the contents. Anything that spilled on to the sand was carefully licked up. As the level of the egg's contents dropped, she broke open more of the shell so that she could get her tongue deeper into the shell.

Once she had eaten the two eggs, she picked up another and carried it away in a north-easterly direction. After 40 metres she stopped and this time carefully placed the egg in a clump of tall grass. She returned to the nest and immediately picked up another egg, carrying this one off in a northerly direction for 600 metres, before putting it down under a bush. Over the next three hours she carried off another 10 eggs in different directions for distances varying between 150 and 600 metres, hiding them under various bushes or in grass clumps. Then she ate another egg at the nest, carried one off for 600 metres, dropped it rather carelessly in a grass clump and lay down close by. At this stage there were eight uneaten eggs in the nest and I left Cicely sleeping and went home to do the same.

The next afternoon when I returned to the nest there were only two broken eggs left. I also saw a male ostrich close by. I had thought it strange that an adult ostrich was not sitting on the nest when Cicely found it. Apparently ostriches do sometimes leave their nests for a short while before starting to incubate. In this instance the birds certainly paid dearly for their negligence.

I soon found Cicely eating an egg and was slightly surprised also to find Floppy-Ears close by, eating one herself. After finishing her egg, Cicely picked up another which she had stored under a bush and carried it away. After walking about two kilometres she put it down in some long grass. Then she moved right away from the nest area. She returned at 2 am, found and ate two more of her stored eggs, and carried a third off for one kilometre, before placing it under a bush. She then moved over to a different bush, where she had stored another egg, ate this, and lay down. Soon afterwards I left her.

This scatter-hoarding of ostrich eggs is one of the most spectacular examples of food-storing behaviour I have heard of. Neither was it the only time I saw this behaviour. On an earlier occasion when we had been following Floppy-Ears, she came to an area and started casting around from bush to bush, as if searching for something. Eventually she produced an ostrich egg from one of the bushes. A short while later she passed a deserted ostrich nest with a few broken pieces of shell. She had obviously stored the egg some time previously and remembered the general area she had stored it in though not the exact bush.

We also observed brown hyenas store other things. If it found a large, meaty carcass, the hyena would quickly break a piece off and carry it away to hide under a bush, in a clump of grass or even down a hole. Then it would return to the carcass and eat from it. If it was dislodged by a larger predator, or if more brown hyenas came to the carcass, it could recover the stored portion. Alternatively, once it had eaten the carcass, it could later return to its stored meal. Normally a hyena will return within 24 hours to recover its bounty, though once a leopard stole a stored steenbok before the hyena returned. A favourite item that brown hyenas store is a piece of dried-out hide. They will return from time to time for a chew, perhaps if they haven't found much to eat lately, sometimes spending as long as an hour, before putting it back in the bush and moving on.

As Normali's three cubs were growing up and were soon due to leave the den, I was anxious to mark them. We set a trap close to the den and scored a hat-trick: three cubs in three nights. We called the two males Shirmile (after our friends Shirley and Emile who happened to be with me when I caught him) and Charlie, and the female Chinki (I am not sure where these two names came from).

In October 1975 we found Ro-Ro, the old female from the Rooikop Clan, in a very bad way near Nossob Camp. She had been bitten around the neck and on the flanks and was very thin. For two days she stayed in the vicinity of the camp, not moving more than a few hundred metres. Elias suggested that we put her out of her misery. Although I do not usually approve of interfering in this way – the powers of recovery of wild animals are amazing – in this case I agreed. Just how desperate she was came home to me when we opened her stomach to find only a few pieces of bone, a large buckle (35 x 45 millimetres), and a few pieces of an old leather belt inside. I had not seen the male Jo-Ro for many months and presumed that he too had died. Luckily I had managed to catch and mark her two cubs, Phiri and Tswana – the Setswana for brown hyena (*phiritswana*) – but for the moment we lost contact with the Rooikop hyenas.

During the latter half of 1975 we also made strides with some data analysis. It was clear that animals that visited the same den inhabited the same territories. We decided to refer to each group as a clan, following Hans Kruuk's term for a group of spotted hyenas. At this stage we knew the composition of six clans, but there was no pattern, so that it was impossible to say what the composition of a typical clan was. The Kaspersdraai Clan only contained the adult female D'Urbyl and her three cubs. The Kwang Clan consisted of a male, Hop-a-Long, and a female, Normali, with three cubs, as did the Cubitje Quap Clan with Gertie and Harken and two cubs. The Rooikop Clan was made up of the male Jo-Ro, the female Ro-Ro, a third adult of unknown sex, plus two cubs, and the Seven Pans Clan had the two adult females, Cicely and Floppy-Ears, plus three sub-adult animals and one cub. A sixth clan, the Botswana Clan, which lived to the east of Nossob Camp, was known to contain two adult males, one adult female, two sub-adults and a cub.

Margie set about mapping all the movements we had recorded from our radio-collared hyenas, a painstaking job at which she was so much better than I. Once this had been done we were able to look at the movement patterns of our hyenas and obtain a measurement of the size of their territories. As with the composition of the clans, the territory sizes of different clans varied. The large Seven Pans Clan's territory covered an area of 460 square kilometres, whereas D'Urbyl's Kaspersdraai territory was only 267 square kilometres. It looked as if the larger groups might have the larger territories. However, our observations of the Kwang Clan over the next few years were to show that the factors controlling both group and territory size are more complicated than a direct relationship between the two and are tied up with food and the way in which it is distributed in the territory.

The year 1976 was even wetter than 1974 had been. Once again the Kalahari became covered with green grass and a myriad of multi-coloured flowers, and the river-beds were invaded by thousands of springbok and hundreds of gemsbok, wildebeest and hartebeest, providing an awesome spectacle. Unfortunately because the roads into the park were impassable except through South West Africa (now Namibia) few people were able to make the long journey to enjoy it.

Early in January 1976, with my repaired receiver, we recaught Normali in order to change her collar as the batteries in the old one had run down. Once again she was in milk. This time when we released her she went straight to her den, which was very close to her last one, and we found two small cubs not more than six weeks old. Hop-a-Long had somehow managed to discard his collar and it was only in March that I finally managed to trap him for the second and last time. In spite of an intensive campaign I was not able to trap either of the hitherto easily trappable Floppy-Ears or Cicely, nor did I ever see them again. I did trap the female from the Botswana Clan and fitted a radio to her, but unfortunately this radio turned out to be a dud and I was never able to follow her. Circumstances, therefore, largely determined that for the first half of 1976 the time I was able to spend with brown hyenas was confined largely to the Kwang Clan. It was an intensive study period during which we followed Normali and Hop-a-Long for a combined distance of 1 235 kilometres.

I was interested to find that Normali's previous litter of cubs, Shirmile, Charlie and Chinki, often visited the new den, where they appeared to be welcome. They often played at the den with the cubs and we were treated to some lovely displays of rough-and-tumble, chasing and pulling, by large and small hyenas together. By April, when they were about 22 months old, Shirmile had disappeared but Charlie and Chinki were still conspicuous.

On most nights that we went out we found one of our two radio-collared hyenas within a couple of hours somewhere along the Nossob river-bed or its environs. One particularly cold night in June, when the temperature had dropped to −5°C, we had been searching for over four hours without picking up a radio signal. I was thinking more and more of a nice warm bed when, almost to my disappointment, I finally received a signal from Normali's collar near Kwang Pan.

Whatever reluctance I had felt to following her was quickly dispelled when we found her on the pan with another, unmarked hyena. As we caught their eye-shine in the spotlight, it looked as if the strange hyena dismounted from her. Normali immediately moved off and the stranger followed her closely. They moved off the pan and up the side of a dune, the stranger now about 20 metres behind her. Normali soon stopped and the stranger came running up to her and immediately climbed on her back. In the long grass it was difficult to see exactly what was happening. We could see that the male was resting his chin on Normali's back and clasping her with his forepaws. After not more than 30 seconds, the male appeared to slip off and the two animals stood side by side for nearly one minute. Then Normali moved off again followed by the male. Over the next hour the male followed Normali closely and mounted her another six times, although we could never be certain that intromission had taken place. Each time he came off, the two animals stood side by side for half a minute before Normali moved on. Frustratingly, our observations came to an abrupt halt when I drove into a hidden aardvark hole and spent the next hour digging out the Land Cruiser. By the time we eventually were mobile again, Normali's radio signal was lost in the dunes.

Nevertheless it had been a most interesting hour. There is no doubt that this was mating behaviour, even if we could not be sure if copulation had taken place. There were of course many unanswered questions: Where was Hop-a-Long, who was surely Normali's mate? Who was the stranger and where did he come from? Although we had seen strange hyenas in the Kwang territory from time to time, we were certain that there was not another resident adult male in the Kwang Clan. We were even more confused a week later, when Willem Ferguson, a researcher who was spending a short time in the Kalahari looking at black-backed jackals, saw Normali mating again on Kwang Pan with a different male. Obviously there was more to the mating system of the brown hyena than a simple monogamous system; this would help explain the varying number of males we had found in the clans.

In July Hans Kruuk visited us again. He was to be the keynote speaker at an Endangered Wildlife Trust symposium in Pretoria and the Trust had very kindly agreed to pay for him to spend some time in the Kalahari. Once again the few days that we spent together in the field were invaluable and very stimulating for me, for although the Kalahari is the most wonderful study area, it is isolated, and I needed to be exposed to the ideas and arguments of other scientists. For a hyena person there was no one better than Hans.

Hans suggested that it was time that I wrote up some of my results. Accordingly I spent the latter half of 1976 writing up my MSc thesis on the diet and foraging behaviour of the brown hyena. Field work for the moment was confined to the odd visit to the Kwang den, to check up on the progress being made by Normali's two cubs, both of which I caught and marked. They were a male and female whom we called respectively Shimi and Sanie, a corruption from the Tsonga name for the brown hyena, *shimisane*, which means 'little hyena'.

CHAPTER 5

Food, Foraging and Foes of Brown Hyenas

By NOW THE BROWN HYENA study had developed into a far larger one than I had originally planned. The potential was there, not only for my master's thesis, but for a doctoral thesis as well. I was beginning to develop ideas and gain some insight into questions regarding the evolution of their social system and behaviour. It was clear that their diet, the way in which they find their food (what scientists call their foraging behaviour) and the manner in which this food is distributed, were crucial in determining the type of society they live in. It therefore seemed a good idea to describe these aspects of the brown hyena's ecology, as well as their relations with other carnivores, for my master's thesis and to use this as a springboard for discussing the more detailed aspects of their behavioural ecology in a doctoral thesis in a few years' time.

We used four methods to study the diet of the brown hyena; tracking spoor, faecal analysis, direct observations and the analysis of food remains at dens. Each has its advantages and limitations. As I have already described, in the early days we invested much time and effort in tracking brown hyena spoor, but this was not very successful. Many of the food items were eaten completely by brown hyenas, leaving no trace when we came along the next day on the spoor. Often we were not able to follow a spoor because the animal had moved into the river-bed or the wind had blown it away in the dunes.

In order to expand our observations of the brown hyena's feeding habits, therefore, we collected faeces (scats) so that we could examine these for the remains of food items. The results from our faecal analysis showed that the brown hyena indeed has a catholic diet. Nevertheless there were several limitations to this method. Most importantly it was impossible to say what proportion of the food eaten by the brown hyenas was scavenged by them and what proportion was killed. Secondly, the frequency with which food items occurred in the scats was not a good indication of their actual contribution to the brown hyena's diet. For example, the fact that beetle remains were found in 35 per cent of the scats, whereas springhare remains were found in only 12 per cent, does not mean that beetles were three times as important as springhares in their diet.

Faecal analysis proved to be really useful in the comparison of the diets of adults and cubs. Of the 383 scats Margie analysed, 240 were from cubs and 143 were from adults. Some food items were far more important in the cubs' diet than the adults' and vice versa. Cubs ate more beetles and termites than adults did, as well as small mammals such as steenbok and bat-eared fox. On the other hand wild fruits and large mammals like gemsbok and wildebeest were eaten far more often by adults. Wild fruits are never carried to the den by adults. Furthermore, large carcasses are rarely carried to the dens, as the cubs' teeth and jaws are not sufficiently developed to deal with the bones, and smaller carcasses are also easier to carry.

It was by following the brown hyenas at night that we obtained the most accurate indication of their diet. We recorded 794 food items that brown hyenas fed on. These included 18 species of mammal – 62 per cent of the total number of mammal species, except for small rodents and bats, recorded from the southern Kalahari. No single item made up more than 10 per cent of the total number of items, the highest in terms of frequency being the two wild fruits, the tsama (9.8 per cent) and the gemsbok cucumber (9.2 per cent). The most important mammals were wildebeest, gemsbok and springbok. However, not even this method was perfect. Often the hyenas ate something small which we were unable to identify. What we were able to verify from the faecal analyses consisted mainly of insects and small reptiles.

The final method we used to study the brown hyena's diet was recording food remains at their dens. This method gives a good picture of the type of food adults bring to their cubs. Over 75 per cent of the 246 different animal remains we found at dens were animals of springbok size or smaller, and nearly 75 per cent of these were springbok, bat-eared fox, black-backed jackal and steenbok. This result tied in with what Margie had found in faeces from cubs. So cub and adult brown hyenas do have slightly different diets. This seems to reduce the competition between these two age classes and makes the comparatively scarce resources in an arid region like the Kalahari go much further.

A striking point about the brown hyena's diet is that nearly all the food items they feed on are comparatively small and provide a meal for only one hyena. Even when they feed on a wildebeest or gemsbok carcass these are often the remnants of kills that have already been fed on by spotted hyenas, lions, vultures and jackals. The significance of this will become clear when I come to compare the behaviour of the brown hyena with the other star of this book, the spotted hyena.

An important question we hoped to answer was what proportion of the brown hyena's diet comes from kills made by the hyenas themselves and what proportion is due to scavenging and other sources. This had always been a rather controversial subject amongst naturalists. For long it had been contended that hyenas, in particular the common and widespread spotted hyena, are plain and simply scavengers – and cowards to boot. However, some of the early naturalists maintained that the brown hyena is an efficient hunter, capable of pulling down animals as large as kudu and zebra. Colonel James Stevenson-Hamilton, the first warden of the Kruger National Park, reported in his book *Wildlife in South Africa* that he once found a brown hyena den along the Sabi River littered with bones and skulls of several animals including impala, baboon and even cheetah. He went on to say that 'it was quite evident that most if not all the animals had been seized alive and killed by these hyenas' although he does not explain why.

I have already suggested that our brown hyenas hunted very little. In fact after following brown hyenas for 3 250 kilometres we saw them make only 12 kills. Top of the list at four were korhaans, which are fairly large ground nesting birds weighing about 700 grams. Second and equal at two each were unidentified 100-gram lark-sized birds and springhares which weigh about 3 kilograms. The remaining animals were a 5-kilogram springbok lamb, a 4-kilogram bat-eared fox, a small unidentified rodent weighing perhaps 300 grams, and Bop's 900-gram striped polecat. I calculated that in terms of kilograms of food eaten, killed prey made up a paltry 5.6 per cent of the brown hyena's food.

Most of the hunting attempts of brown hyenas are unsophisticated. The majority of the 151 attempts we saw took place when the hyena chanced upon a small animal and gave chase. Most were after springhares, bat-eared foxes and springbok lambs. Springhares frequently escaped down a hole from which the brown hyena had no chance of digging them out. Bat-eared foxes often co-operated in defending themselves. If a hyena gave chase after a fox, any other foxes in the area were attracted by the disturbance. They would run over to the hyena and cut in front of it with their tails held aloft and curled forward in an inverted U-position. Then when the hyena abandoned the hunt, the foxes would mob it by following it closely, jinking around it, sometimes uttering a high-pitched bark, and causing the hyena to scuttle off with its hair raised and tail curled under, looking more like a scolded wretch than an aggressive hunter.

For most of the year brown hyenas ignore springbok, but during the lambing season, when the springbok often congregate on lambing grounds, their behaviour changes. This is the only time they showed any kind of formality in their hunting behaviour, although even then the results were unimpressive. One bright moonlit night we followed Normali on the limestone plains south of Kwang Pan when there were several hundred springbok with lambs scattered in the vicinity. Normali was moving erratically, with her nose to the ground much of the time, obviously looking for lambs. Suddenly a lamb jumped up two metres from her and ran off. She lunged at it, but did not chase it. A little further on she flushed out a second lamb and chased it for 50 metres, but the lamb easily escaped. Then she practically stumbled on a lamb which ran off, but again she obviously realised that she had no chance of catching it and only made a reflex lunge. The next lamb she encountered ran off and she chased hard for 300 metres, but the lamb was too fast for her. After standing for several minutes, looking and listening, she continued moving through the lambing grounds, nose to the ground. Further on she made a sharp turn into the slight breeze, moved at a fast walk and then ran for 15 metres after a lamb which bounced away, in the *'pronking'* gait which springbok often display towards predators. She came up to the place where the lamb had been lying and sniffed at it briefly. After spending nearly three hours amongst the springbok she eventually moved away after making several more unsuccessful attempts to catch one.

Even though Normali was actively trying to hunt springbok lambs, her attempts were rather clumsy. On the one occasion that we saw her catch a lamb, she flushed it out of some tall grass and chased it for just over a kilometre at a speed of about 30 kilometres per hour, easily the most impressive hunting effort by a brown hyena we saw. It is interesting that she chased this one for such a long distance, whereas usually she gave up after 50 or so metres. Did she detect some slight physical defect in this lamb and therefore put in a special effort because her chances of success were better than normal?

Two species of korhaan, the northern black and the red-crested, are widespread in the dunes of the Kalahari. They were the birds most frequently hunted by our brown hyenas. Although we only saw 10 korhaan hunts, five of them resulted in a kill, a relatively high proportion. At night the birds roost on the ground in tall grass and it is only if a brown hyena passes really close that it will detect a bird. Then it will stop, turn abruptly towards the korhaan and move at a fast walk with its head down. On five occasions the korhaan flew off in time, in four it did not and once the korhaan flew off, but the hyena ate two small chicks.

Our observations clearly showed that the brown hyena is a poor hunter, capable of pulling down only small prey and then as an exception, rather than the rule. Studies of the brown hyena in other areas have confirmed this. In the central Kalahari the Owens also found that scavenged vertebrates were the most important source of the brown hyena's food and that they hardly ever hunted. Glyn Maude found that scavenged zebra in and around Makgadikgadi National Park, Botswana, and scavenged livestock outside the park were the brown hyena's chief food, although they occasionally caught hares and springhares. The one slight exception to this rule occurs on the Namib Desert coast. Here Ingrid Wiesel has for many years studied the brown hyena, where it is colloquially known as the *strandwolf* (beachwolf). These brown hyenas often move up and down the coastline where they search for carrion washed up along the beach. Cape fur seal breeding colonies are food hotspots. Here both live and dead seal pups are available throughout the year, but especially during the seal pupping season between November and January. At this time brown hyenas prefer to kill the small, immobile pups even though there are also quite a few dead ones available from starvation. An interesting fact noted by Ingrid was that although the brown hyenas find it easier to kill small pups, they are more likely to eat the occasional larger one they manage to secure and often leave the small pups uneaten. It is one of the few occasions that brown hyenas can afford to be a bit fussy.

When you look at a brown hyena with its weakly developed hind legs and long, shaggy coat, it certainly doesn't give the impression of being built for speed or stamina, features that would surely be needed in order to overcome such formidable prey as zebra, kudu and cheetah. So where do the stories of their hunting prowess originate? In 1974 we spent some time in the Kruger National Park investigating the current status of the brown hyena there. As I will discuss later, I believe that there is no breeding population of brown hyenas in the park today, but that is not of relevance here. During our investigations I was able to question many of the rangers and game guards about the brown hyena. I found that, almost without exception, there was confusion amongst the game guards in Kruger between the very rare brown hyena and the abundant spotted hyena. None of them were really sure of the difference in the spoor between the two species. They also were in agreement that the spotted hyena is not able to hunt anything larger than an impala lamb, whereas the brown hyena is a hunter of large prey. We now know of course that the spotted hyena is indeed an efficient hunter. Granted the amount of hunting it does varies from area to area, and in Kruger it is not as predacious as in some other areas, but I think it likely that when the game guards in Kruger came across a carcass that had obviously been killed by hyenas, they incorrectly assumed that it had been killed by the brown hyena.

Stevenson-Hamilton and other naturalists were, no doubt, also influenced by the local people. They incorrectly reasoned that food remains found at brown hyena dens were proof that the hyenas had killed them. Spotted hyenas rarely carry food back to their dens, so they do not become littered with prey remains. If they did, perhaps the early naturalists would also have believed that the spotted hyena is a hunter. I firmly believe that the only zebra or kudu a brown hyena will pull down is either a very weak foal or calf, or appears in its dreams!

Perhaps the most surprising finding from our study of the brown hyena's diet was the very high incidence of wild fruits. Incongruous sights were those of brown hyenas picking the small berries off brandy bushes or from ground runners of an annual ground creeper called merremia. We even saw Cicely eating Kalahari truffles. These fungi grow just under the ground in calcareous soil at the end of the rainy season. Cicely, like pigs in France, was apparently able to sniff out these highly nutritious plants. Human inhabitants of the Kalahari relish these truffles, which to me tasted very ordinary, a bit like mushrooms but not as tasty. However, the ones we ate were not prepared with any additional herbs and spices. Perhaps our western-influenced palates demand the garnishing that usually goes with the preparation of this type of food.

But it was the tsama melon and the gemsbok cucumber that were the really important fruits in their diet. The smooth-skinned tsama is an annual fruit with an edible portion of 300 to 700 grams. The oval and spiny-skinned gemsbok cucumber is a perennial with the edible flesh weighing 100 to 200 grams. Both grow from ground runners, in patches, mainly in the dunes, and become available from about March each year. However, their abundance varies from year to year, depending on rainfall. Being an annual, the tsama is the more fickle of the two, so that in some years there are effectively none, whereas in others we counted densities of over 1 200 per hectare in some areas. Tsamas are resistant to frost and can remain edible for over a year. Gemsbok cucumbers, on the other hand, are susceptible to frost and only last until mid-winter.

Although eaten with such relish, the tsama and gemsbok cucumber have a low nutritional value. I calculated that a brown hyena would have to eat 22 tsamas to obtain the equivalent energy of one kilogram of fresh meat. The nearest a brown hyena got to this was one of the Cubitje Quap cubs who ate 18 in a night. They are, however, rich in trace elements and Vitamin C. The fruits also have a very high moisture content of over 90 per cent and are obviously an important source of moisture for many of the Kalahari's inhabitants, including the San. For brown hyenas, which rarely eat fresh meat and therefore obtain less moisture from their diet than do the other carnivores, this might be important. Unfortunately, I was unable to establish how important they were as a source of moisture to brown hyenas, as all the hyenas we studied had access to water from boreholes. We did find that the number of fruits they ate per night was not dependent on whether they had drunk that night or not, which does suggest that the nutrition gained from tsamas and cucumbers was important. Once a brown hyena knows where the tsama and gemsbok cucumber patches are in its territory, it does have an easily accessible food source to tap while stocks last.

Tsamas and gemsbok cucumbers are also eaten by many other animals. Their main nutritional value is in the seeds and these are highly fancied by rodents. Brown hyenas open the fruits up and make the seeds available to small rodents. More importantly, seeds that are ingested by the hyenas pass through the digestive system in the faeces, where they are protected from rodent predators and thus have a chance to germinate. Brown hyenas may be important agents in the dispersal of the seeds of these wild fruits.

Brown hyenas, like all carnivores, are capable of eating large quantities of food quickly, should the opportunity arise. Cicely ate 7 ostrich eggs, equivalent to 168 chicken eggs, in two nights! By the same token they are able to operate quite efficiently on almost nothing for a surprisingly long time. D'Urbyl once stayed away from her four-month-old cubs for five nights, without their suffering any ill effects. It has been described as a system of feast or famine.

But what is the average daily consumption rate of southern Kalahari brown hyenas? In an attempt to find out, we followed Normali and Cicely, both of whom were lactating at the time, for five periods of two nights and once we followed Ro-Ro for three consecutive nights. During these periods we attempted to estimate the weight of all the food items they ate. We collected a sample of bones of various sizes, ostrich eggs, and the lizards and beetles we saw them eat, and weighed them. For mammals that were eaten we looked up the mass of the particular animal in references, gauged what proportion the hyena ate, and converted this into a measurement. In the 13 nights that they were followed the hyenas ate a total of 138 food items which we calculated to weigh 36.5 kilograms. This averages out at 2.8 kilograms per hyena a day.

Not surprisingly, brown hyenas and their two closest living relatives, the spotted and striped hyenas, have evolved several unique adaptations to a scavenging life. They have extremely powerful jaws and teeth for crushing bones – they are the only carnivores capable of breaking into the leg bones of large animals like wildebeest. Locked up in these bones is highly nutritious marrow, often rich in fat. They are also able to completely digest the organic matter in bone and obtain the maximum available energy, another scavenging feat that no other carnivore can perform. The chalky white faeces characteristic of hyenas contain the indigestible remains of bones they have chewed.

Another useful adaptation that hyenas have evolved in order to help find carrion is an uncanny sense of smell. When foraging, brown hyenas continually monitor for the presence of carrion by sniffing the air. When a breeze is blowing the hyena repeatedly lifts its head into the wind and sniffs. When there is no breeze blowing they also sniff the air repeatedly, but then the sniffs are in no particular direction. Under windless conditions they often have difficulty in pin-pointing a smell, but it was amazing to see how they were able to detect smells of even dried-out old carcasses from over two kilometres, provided they were down wind from them. On the other hand, we once watched Ro-Ro pass within 10 metres of a tree in which a fresh springbok carcass was hanging. There was a strong wind blowing from her to the carcass and she failed to smell it.

Brown hyenas are also able to follow scent trails. Cicely obviously followed a scent trail to the ostrich nest and on another occasion we followed Normali on a scent trail for 1.5 kilometres to a springbok carcass. We were able to use this faculty of hyenas in our trapping programme and frequently dragged the bait for one or two kilometres on either side of the trap before setting it. This improved our trapping success.

Hearing is also well developed in hyenas. Brown hyenas respond to distress calls of antelope and springhares by running off in that direction should they hear one. The most spectacular example I have observed of the use of hearing involved a spotted hyena. The animal had been lying down for an hour and a half when it suddenly stood up, turned with its ears cocked in a particular direction for a few seconds, then ran off in pursuit. It kept running, then stopping and listening briefly, for an unbelievable 10 kilometres until it arrived at a carcass on which other hyenas from its clan were feeding. It had obviously heard these animals squabbling over food at the carcass.

Sight plays very little role in scavenging, being more important for a hunter. Nevertheless brown hyenas have good night vision, far superior to ours, although I am inclined to think that they do not see so well during daylight.

Having established that brown hyenas are successful scavengers, we obviously wanted to find out where this scavenged material originated. Many of the scavenged items we saw our hyenas eat were odd pieces of skin and bone which seemed to be widely scattered throughout our study area, and we were unable to trace their origin. However, when they scavenged from a fresh carcass, we were usually able to establish how it died. Nearly half (42.7 per cent) of the feeding occasions involved carcasses from lion kills. Typically, if a brown hyena found a carcass on which lions were feeding, it would circle round at a safe distance, watch the lions intensely for a few minutes, then move off, returning from time to time to check how the lions were getting on, until they finally departed. The hyenas were left with little meat but with a good selection of fresh bone and skin. Lions, therefore, were generally of benefit to brown hyenas.

Sometimes, however, lions may be a threat to brown hyenas and attack them for no apparent reason. Once when we were following Hop-a-Long down a road, a lioness jumped out at him and chased him a good 50 metres. Sometimes if a brown hyena lingered too much near a carcass the lions would give chase for a short distance. The most spectacular interaction between brown hyenas and lions was observed by a group of students from the University of Pretoria. As part of a field course the students were conducting a 24-hour game count at Cubitje Quap Windmill. Peter Apps recorded the sequence of events in the *South African Journal of Zoology* as follows:

'A loud, drawn-out call, somewhere between a howl and a scream, was heard from the direction of the wind pump. We drove up to investigate and found a pride of two lionesses and five sub-adults 30–40 metres from the nearest cover. One of the lionesses was carrying a brown hyena by the throat. After walking for 10 metres she dropped the hyena which lay motionless as the rest of the pride gathered around it. The hyena lay motionless for 1–2 minutes, then kicked spasmodically. The lioness immediately straddled it and again bit its throat, holding on for 3–4 minutes. When it was released the hyena dropped motionless to the ground, appearing to be dead except for the occasional blinking of its eyes which still reflected the spotlight beam. Since it remained motionless for 10 minutes while the lion pride moved away some 10 metres and lay in a group, I assumed that the hyena had been suffocated into unconsciousness by the lioness's throat bite.'

Later events were to suggest that, on the contrary, the hyena was shamming rather than unconscious.

'The hyena was still motionless 10 minutes later when one of the sub-adult lions returned to it, rolled it over with a forepaw and wrestled with it like a domestic cat playing with a large rat. As it was thrown into the air the hyena either recovered consciousness or stopped shamming, rolled onto its feet and faced the lion with the same scream heard earlier. The lion was deterred long enough for the hyena to turn and run towards the dunes, limping heavily on one foreleg. The whole lion pride pursued it, though apparently rather half-heartedly because their stomachs were full from feeding on an adult blue wildebeest which they had killed the previous night. The hyena reached the sparse bush cover on the dunes well ahead of the lions, who abandoned the chase and returned to rest near the wind pump.'

Peter Apps concluded from this observation that the brown hyena was shamming death in an attempt to avoid attracting further attention from the lions. He points out that shamming appears to be found in aggressive encounters between predator species when the larger combatant is motivated by aggression rather than by hunger, probably because it is the only successful way for the attacked animal to prevent the larger one killing or severely injuring it. I have witnessed similar behaviour in a jackal which was attacked by a leopard. The jackal allowed itself to be carried several hundred metres by the leopard, by lying limp in its mouth. Eventually the leopard put the jackal down under a bush and moved off a short distance. When we arrived some minutes later the jackal ran off, shaken but able to survive until another day. Peter Apps concluded his report from which I have quoted by saying that he saw the hyena three days later, although its long-term survival was unknown and it was still weak and lame from its injuries.

After lions the next most important agent in providing carcasses for brown hyenas was what I call non-violent death; that is, death due to starvation or disease. When we could ascertain the cause of death, nearly a third (29.9 per cent) of all carcasses on which brown hyenas fed showed the effects of this agent. Nearly all of them were wildebeest which died during the years of drought between 1978 and 1980. These carcasses were different from the lion-killed carcasses in that they often contained a good proportion of meat, not having been eaten by spotted hyenas or lions.

Brown hyenas rarely scavenged from leopards, because leopards have the habit of dragging their carcasses into trees, thus removing them from their competitors. Several times a brown hyena came to a tree in which a carcass was hanging and spent several minutes circling it and looking up longingly at the carcass. The best the hyena could do was to pick up a few scraps from under the tree. On the few occasions that we saw encounters between brown hyenas and leopards, they tended to ignore each other, although I once found the remains of a six-month-old brown hyena cub that had been eaten by a leopard.

Cheetahs are known for their lowly status in the predator hierarchy. Because of their rather small heads and teeth, which are adaptations for speed, not for fighting, they do not quarrel with their larger competitors over food. Their generally solitary nature also means that they dare not risk injury over a food dispute which could render them ineffective hunters.

Not long after we arrived in the Kalahari we found a cheetah female with three small cubs on a very freshly killed springbok. The mother was exhausted after the hunt and lay panting next to her kill for over half an hour. When she finally started feeding and opened up the carcass for her cubs, it was almost dark. They hadn't been feeding 10 minutes when the mother suddenly sat up and looked off intently at something. Then we saw a large brown hyena standing close by. The cheetah ran at the hyena, while her cubs scuttled off in the opposite direction. The hyena retreated a few metres, then turned round and, with its mane and tail erect, charged at the cheetah. She stood her ground and they clashed briefly, the cat slapping and growling at the hyena. The hyena was not intimidated and kept moving forward, at which the cheetah spun round and ran off to her cubs, abandoning her hard-won meal to the hyena.

Such incidents are rare in the southern Kalahari as cheetahs are not very common. However, along the Auob river-bed there was a higher density of cheetahs than in other areas. I often found cheetahs feeding on springbok kills along the Auob, and whenever I checked the carcass the next day it had been removed by a brown hyena. From this it appeared that a few brown hyenas derive much benefit from cheetahs.

Two smaller carnivores have an important relationship with the brown hyena. The caracal is comparatively rare, yet we saw brown hyenas steal kills from one on eight occasions. Each time the hyena merely came running up to the caracal, which gave way without an argument. The kills were three steenbok, two springhares, and one each of a springbok adult, a springbok lamb and an African wild cat. All these carcasses provided the hyenas with a good meaty meal.

Brown hyenas also sometimes steal kills from the ubiquitous black-backed jackal. We saw them take four steenbok, a springbok lamb and a springhare. We also saw on three occasions a brown hyena dig out a dead rodent which had probably been buried by a jackal. Furthermore, as our observations of food remains at dens in particular showed, brown hyenas regularly eat jackals, although we have no evidence that the hyenas kill them.

At other times the shoe is on the other foot and jackals deprive brown hyenas of food. Jackals are often quick to find a carcass or get on to it first once lions or other carnivores have left. They are deceptive in the speed with which they eat the meat. Their razor-sharp teeth are able to open up even large carcasses such as a wildebeest, and they bore into the meat like mice into a piece of cheese. When they have filled their stomachs they move away, dig a hole and regurgitate into it, before covering it up. They are then ready to start feeding again. By the time the first brown hyena comes along most of the meat may be gone.

Jackals can also be a great nuisance to feeding brown (and spotted) hyenas. It was often noticeable when a brown hyena started to feed on bones how one or two jackals were attracted to it. I think that the sound of bones being cracked alerted jackals to this fact. In this situation the jackals come right up to the hyena and, by persistent pestering, sometimes to the extent of nipping the hyena on its heels, cause the hyena to move away a few metres. Immediately the jackal will go to the spot where the hyena was feeding and pick up the scraps, which, if left alone, the hyena would have eaten itself. The hyenas often retaliate by lunging at the jackals, but it is a futile exercise. The cunning jackals are too quick.

The most flustered brown hyena I ever saw was the young male Charlie from the Kwang Clan. We found him one night near Kwang Pan carrying the best part of a large springbok carcass in the direction of the den. With the legs hanging down and the head and horns to contend with, the last thing he needed was the attention of the six jackals that were following him, nipping him in the back legs, trying, it seemed, to make him drop the food and turn on his tormentors. With hair raised he tried to run away. This only aggravated the situation, as he then tripped over his cumbersome load and dropped it. He then had to fight off the jackals while trying to get a good grip on the carcass so that he could carry it off again. It took Charlie nearly half an hour before he was finally able to accomplish this and rid himself of the jackals. He didn't lose any meat but I am sure he lost his composure.

The last meat-eaters that compete with hyenas to any significant degree in the southern Kalahari are vultures. White-backed vultures are common here and are efficient scavengers, able to cover large areas quickly in their visual search for food, and rapidly dispose of meat at a carcass by sheer weight of numbers. Vultures are active by day, whereas hyenas are mainly active at night. One of the reasons for hyenas taking up a nocturnal existence may be to lessen competition with vultures. Because vultures are better at finding carcasses during the day than hyenas are, it seems sensible for hyenas, with their very good sense of smell, to search for carrion at night. Additionally, the hyenas' nocturnal behaviour is an important way in which they save water. If they moved around a lot in the hot African sun, they would have to use up water by panting in order to keep cool. Obviously over much of Africa this would be a poor strategy.

Vultures are dependent on hyenas in an important way. Because they are unable to chew bones, they depend on hyenas for bone fragments to supply their chicks with calcium, essential for bone growth. Vultures apparently return to carcasses after hyenas have fed in order to obtain the bone fragments. In areas where hyenas have been exterminated it has been found that vulture chicks suffer from bone deformities. Researchers have discovered all sorts of white artefacts such as pieces of glass and porcelain at these nests, which have obviously been picked up by the vulture parents in the belief that they were pieces of bone. Recently conservationists have erected a series of vulture restaurants in agricultural areas where carcasses are provided to feed vultures, and bones are smashed up in order to supply them with calcium.

Although the brown hyena is very much a scavenger, this does not mean that it has an easy life living off the pickings of others. Carrion is thinly and widely scattered in the desert, and the brown hyenas have to work hard in order to find it. They usually become active at about sunset and very often don't rest again until the sun rises. We found that on average they were active for a staggering 80.2 per cent of the night, during which time they would move an average of 31.1 kilometres. The record for one night was held by Floppy-Ears at 54.4 kilometres. Recently I read an article on the ranging behaviour of caribou in Alaska. The authors calculated that over a year these migratory deer cover distances of some 5 000 kilometres and that these were the longest movements documented for any terrestrial mammal. Kalahari brown hyenas might not range over such large areas as do Alaskan caribou, but by moving approximately 11 350 kilometres a year, they cover over twice the caribou's distance!

This perpetual motion of the brown hyenas made it very tiring following them for long periods. Some nights we could not even stop for a cup of coffee, until we purchased some baby plastic drinking cups. This made it possible to enjoy some refreshment without running the risk of spilling boiling water in our laps. Nonetheless some nights were very tiring and the effort to stay awake and alert became a real battle. One particularly bad night would not end as Charlie refused to lie down when the sun came up – he just kept moving on. In my exhaustion I became almost hallucinatory and decided that the only way to stop this was to run over the hyena. Fortunately sanity prevailed and eventually Charlie decided that he too had had enough and lay down.

Brown hyenas are quite choosy about the place to rest for the day, particularly in the heat of summer. It was most unusual for any of our study animals to spend the day along the Nossob river-bed; almost always they slept in the dunes. Perhaps they feel exposed to the attentions of larger and more aggressive carnivores in this very open habitat. In summer they either spend the day down a hole or under a large, shady shepherd's tree. Before resting for the day they examine several trees until they find one to their liking. It is important that the tree has thick branches close to the ground, so that the hyena can crawl well under it for unbroken shade. Once the hyena has chosen its resting place it digs itself a hollow in which to lie. During the course of the day the animal will repeatedly kick sand onto its stomach. The temperature of the sand even a few centimetres below the surface is as much as 10°C lower than on the surface, so this behaviour helps with cooling.

If you or I tried to sleep under a shepherd's tree during the day we wouldn't last five minutes, for in the sand under these trees live hundreds of little tick-like creatures called sand tampans. They are blood suckers. They react to carbon dioxide exhaled by animals. Within minutes of an animal taking shelter under a tree, the tampans come popping out of the sand looking for a meal. Wild animals are apparently not worried by sand tampans, but people and even domestic animals may develop painful, suppurating ulcers from the bites, which can on occasion weaken them to the point of death. The mouth of a sand tampan is said to exude an anaesthetic, so its host is unaware of its bites and does not try to rub it off. In my experience the bites of the tampans are painful, and it is these that make it impossible for a person to lie under a shepherd's tree without some form of protection.

In winter, keeping cool during the day is not a problem for a brown hyena. At this time of the year the hyenas use a small bush or even a clump of thick, tall grass as a daily resting site. I never knew of an adult brown hyena sleeping down a hole or under a shady tree during winter.

An important feature of a carnivore's foraging behaviour is the size of the group it forages in. This has consequences for the social behaviour of the species. Brown hyenas are almost exclusively solitary foragers – over 99 per cent of our observations were of single animals. Brown hyenas scavenge most of their food, which mainly consists of small food items. There is no advantage in foraging in a group, as the chances of finding carrion are not improved – one hyena as is good as several when it comes to smelling carrion. But more importantly, when food is found there is usually only enough to feed one animal. If the brown hyenas were in a group it would lead to unnecessary aggression and waste of energy fighting over the food, and in the end result one of the animals, usually the strongest, would take everything.

This point was vividly illustrated to us one night when we found Charlie and Chinki moving together. It soon became apparent that the two sub-adults were actually more interested in playing than in looking for food. Suddenly they flushed out a springhare. Being opportunists, like all carnivores, they began to chase the springhare. Chinki saw it first and was about 10 metres ahead of Charlie. The chase went round in a large semicircle, when suddenly the springhare doubled back past Chinki, almost right into Charlie, who snatched up the prize. Immediately Chinki was after him. Once again the chase was circular, until after several minutes they stopped at a large fallen tree, Charlie on one side, Chinki on the other. The two stood and looked at each other, panting heavily. Then Charlie moved off slowly holding his tail up in victory. After a short distance he put the springhare down and stood over it, still very out of breath. Chinki approached briefly, but then moved off. She was clearly too exhausted to pursue him any further and left him to eat the springhare on his own. It took Charlie a good half-hour to recover fully from all the running around. In fact he even ate the springhare lying down, something brown hyenas rarely do. I have no doubt that the fact that two of them chased the springhare contributed to the success of the hunt. However, the prize was too small to be shared and it was not worth the effort for Chinki to continue trying to get Charlie to share it.

Although brown hyenas are so blatantly solitary when they forage, we came to realise out that they are in fact secretly social. Although a female with cubs at the den is the basic social unit, as we had discovered in some territories other hyenas share the area and even co-operate in feeding cubs. Over the next few years our observations of the Kwang Clan were to teach us much about brown hyena society and how it operates.

CHAPTER 6

The Social Life of Brown Hyenas: Blatantly Solitary, Secretly Social

My MSc thesis was written up by February 1977, but it was only in July of that year that I was able to spend much time with the hyenas again. Other duties as research officer, as well as a wonderful tour of the US (courtesy of the National Parks Board) on which Margie accompanied me, occupied most of my time until then. Butch Smuts, who was working on lions in the Kruger National Park, and I had been invited to present papers on our work at an international symposium on carnivores in Seattle, and we also attended the 13th International Congress of Game Biologists in Atlanta. Additionally, we were fortunate to be able to visit some beautiful and interesting areas, such as the Grand Canyon, Yosemite National Park, the Kenai Peninsula in Alaska, and the Superior Forest in Minnesota. Here we were privileged to visit the famous wolf biologist, David Mech, and to see some of these elusive animals from the air.

By this time the batteries in all our radio collars had run down, so we had to start trapping again in order to fit new collars. Normali was the priority as we were anxious to see if she had bred again and, if so, to find her den. However, the first animal I trapped was Normali's son Charlie. This was a bit of a surprise because he was now three years old, and I would have expected him to have left his natal clan.

I thought it worthwhile to give him a collar and we followed him the night he was released. It was not a very auspicious start as after about two hours we lost him. The radio signal became fainter and fainter as we drove around frantically in the dunes trying to regain visual contact, only to lose the signal altogether. Perhaps we were a bit rusty, not having followed hyenas for almost a year.

Charlie had been moving in a north-easterly direction from Cubitje Quap in the direction of the old dens. Though it was a bit of a long shot we decided to drive out there and try for a radio signal. Once again we were surprised when we picked up a strong signal from his radio transmitter as we came into the area of the dens. This time we had little trouble homing in on the signal. As it became louder I started scanning with the spotlight in order to find Charlie. Almost immediately we picked up the reflections of several pairs of eyes, which at first I thought were those of black-backed jackals around a kill. This notion was short lived as I drove closer and made out the unmistakable profile of brown hyenas. The next instant they ran into a hole and only Charlie was left standing on the mound. He was obviously at Normali's new den and she apparently had a large litter. Five, or maybe six, cubs had disappeared down the hole.

We settled down to wait some distance away. It was a dark night, so we used the spotlight intermittently to check the den. Charlie was lying peacefully, his head pointing towards the hole. Within a few minutes a cub's head appeared. We shone the spotlight to one side, so that it cast only a shadow on the den. The cub came out slowly, its hair raised, not too sure of itself. Perhaps the fact that Charlie was so relaxed gave the cub confidence, for it soon came right out of the den and stood on the mound looking at us. Then a second cub appeared and a third. The first two cubs were quite large, about nine months old, but the third was much smaller, probably not more than four months of age. Soon one more large and two more small cubs emerged from the den. Here was something very different from what we had observed at all the other dens – there were two litters of cubs here.

Our first thought was that Normali had given birth to two litters very close together, but that seemed improbable. More likely, two females were breeding at the den. If so, who was the other mother? We didn't have to wait long to find out. Just before sunrise Chinki, Charlie's litter mate, arrived at the den. The three small cubs came

69

running over to her beg-calling and she lay down to suckle them. The larger cubs were also quite interested in drinking, but Chinki repelled them.

We soon discovered that there were now 12 members of the Kwang Clan, by far the largest brown hyena clan we had seen. Besides the two breeding females and Charlie, there were Normali's previous cubs, the male Shimi and the female Sanie, who were about 18 months of age, the six present cubs, and old Hop-a-Long. Originally we had thought that Hop-a-Long was Normali's mate, but in view of the mating activities of Normali the previous year, this seemed unlikely. Perhaps he was her uncle or some sort of cousin.

We were interested to see who was feeding the cubs. Would the mothers suckle only their own cubs or would they, like lions, not be too fussy about which cubs they gave their milk to? Whenever one of the mothers came to the den, all the cubs would follow her and attempt to suckle, but the cubs that were not her own were usually less persistent. If they did persist, the adult would repel them quite gently but firmly. Then one night, about six weeks after we had found the den, Chinki, who was now also wearing a radio collar, came to the den. Two of her cubs ran up to her, beg-calling, and she immediately lay down on the mound to suckle them. After a few seconds one of Normali's cubs approached, settled down with the others without any opposition, and started drinking as well. After six minutes, Chinki's third cub appeared. He tried to join in but was prevented from gaining access to a teat by Normali's larger cub. Though the two were having quite a tussle, Chinki and the other cubs took no notice. The smaller cub managed to dislodge the larger one but was unable to drink properly as the dislodged cub kept on trying to win the teat back. Eventually Chinki intervened and snapped at Normali's cub, which then moved away.

On other occasions at this den we saw Chinki twice and Normali once suckle cubs that were not their own, out of a total of 15 instances we observed of suckling. Allosuckling, as this behaviour is called, is therefore occasionally practised by brown hyenas, although females show a clear preference for their own offspring. This behaviour provides a good example of an important biological concept known as kin selection. Much animal behaviour can be interpreted according to the hypothesis that the driving force of evolution is the need for animals to pass on their genes. The most common way of passing on genes is through reproduction. By definition offspring share 50 per cent of their genes with each of their parents. However, because Normali was Chinki's mother, Chinki was also closely related to Normali's present litter of cubs; she was, in fact, their half-sister and therefore shared 25 per cent of their genes, assuming (as I have) that each litter of cubs had a different father. Similarly Chinki's cubs shared 25 per cent of their genes with Normali. (Normali's and Chinki's cubs, incidentally, shared 12.5 per cent of their genes, which made them as closely related to each other as first cousins.)

Of course, the fact that parents are most closely related to their own offspring explains why they give preference to them. We can also understand why animals may help other relatives survive by, for example, allowing them to suckle, as some of their genes are carried in those cubs. The amount that animals invest in others is a complicated balance between their degree of relatedness, the energetic costs involved and the chances of reciprocation. It is a fascinating area of animal behaviour that has caught the attention of many zoologists.

Not long after we witnessed the first bout of allosuckling, we made another exciting discovery as far as kin selection and feeding cubs are concerned. This time the main actor was Charlie. We had followed him the previous night and returned to his daytime resting-site early in the evening to continue our observations. It was always satisfying to start an evening's observations with an animal in the hand, rather than having to go out and find one first. At 5.30 pm he emerged from the bush under which he had spent the day, without so much as a glance at the Land Cruiser which I had parked 20 metres away. That was also satisfying – to be one of the very few people lucky enough ever to watch these wonderful animals at close quarters, totally relaxed and unconcerned by our presence. This is something that money can't buy.

Once he was out in the open, Charlie proceeded to groom himself thoroughly, just like a domestic cat, scratching, licking and 'flea biting' himself all over. Then he lay down for a final 'hyena nap' before starting the serious business of finding food. The last winter rays of the setting sun reflected off his black fur as if it had been recently polished, and his white ruff took on a lovely orange sheen. After 10 minutes he stood up, yawned and stretched, moved away a short distance to defecate, and off he went. With the Land Cruiser in second gear and low range we followed.

After going along quietly for 45 minutes, Charlie stopped momentarily, his ears cocked, his eyes focused on something ahead. Then he ran over to a small shepherd's tree from under which we glimpsed a caracal scuttle off into the dark. Charlie came up to the tree and immediately started eating something. We drove closer but could not see what it was. After feeding for several minutes, he picked up the carcass and turned round with it, giving us the view we wanted. What a surprise! In his mouth was an African wild cat. This was a most unexpected prey for a caracal, although subsequently we have several times found one eating its smaller relative. In fact, a number of studies have recently revealed that among carnivores, competition between closely related species is often intense with the larger species dominating; for example lions kill cheetahs, wolves kill coyotes and black-backed jackals kill Cape foxes.

Charlie then moved off carrying the remains of the cat. Was he going to carry it to a more suitable place to eat or was he going to store it somewhere? After he had moved steadily in one direction for some distance, a third possibility dawned on us. Was he going to carry it back to the den for his mother's and sister's cubs? Sure enough, after carrying the wild cat carcass for 3.4 kilometres he came to the den.

At this time the den consisted of two holes some 100 metres apart. Charlie came to the south hole and stood peering in for a minute with the food still in his mouth. Then he put it down and lay down himself. After 15 minutes no cubs had appeared, so he picked up the cat and carried it to the north den. A large and a small cub were lying there and Charlie dropped the food in front of them. The cubs came over to him and sniffed him as he dropped onto his knees and flattened his ears out sideways. This was surprising, as it is the submissive posture usually taken up by cubs when greeting adults, not the other way round. The cubs then picked up the food Charlie had brought and, tugging at it in different directions, disappeared down the den with it. A few minutes later the larger cub came out of the den and ran off with a small piece of cat. It was followed by a small cub, which went up to Charlie, cutting in front of him, beg-calling and rolling on to its back in absolute submission. We later discovered that the other two large cubs and the third small cub were away from the den at the time, so only one cub was down in the den, presumably feeding on the wild cat.

This was a typical sequence of events when food was brought to a den. Though the cubs might squabble over the food at first, they quickly carried it underground. They then took turns, one at a time, feeding. I was never able to establish how they sorted out the order in which they fed: there didn't appear to be a dominance hierarchy amongst the cubs.

This then was the first time that we had recorded a male hyena feeding cubs who were definitely not his own. Charlie had taken on the role of helper at the den. Subsequently we were to learn that male brown hyenas sometimes stay in their natal clan for many years, some like Hop-a-Long probably for life. These males never showed any sexual interest in the females in their group, nor, as far as I could tell, did they ever visit females from neighbouring clans and try to mate with them. They apparently choose a life of celibacy, during which they help animals closely related to them survive, thus exhibiting kin selection.

Brown hyenas generally live in a fairly egalitarian society. We found very little evidence for a dominance hierarchy among the members of the Kwang Clan and rarely saw any aggression. However, during the latter half of 1977 Chinki's relationship with her younger half-sister, Sanie, born to Normali towards the end of 1975, appeared to deteriorate. The first inkling of this was gained one morning in August when we arrived at the den with Chinki. Both Shimi and Sanie were there and Chinki stood and stared at Sanie, who, surprisingly, went down into the den. Chinki then lay down to suckle two of her cubs. After being suckled, the cubs moved a short distance away and started muzzle-wrestling, soon being joined by Sanie, who in the meantime had come out of the den. This seemed to trigger an aggressive response from Chinki. She immediately moved over to Sanie, grabbed her by the side of the neck just behind the ear, and proceeded to pull her around quite roughly. Sanie whined softly in protest and Chinki, her hair raised, replied by growling softly. Chinki pulled her into the den and out of sight. Loud growls and soft yells emerged from there, followed by silence. After seven minutes Sanie came running out of the den, pursued by a determined-looking Chinki. They ran over to a bush and the chase continued around it. Every now and then they stopped and looked at each other, then the chase started again. Eventually Sanie made a break from the bush, but Chinki chased after her. They came to another bush, and again the chase continued round and round, although by now they had slowed down somewhat. Eventually Chinki lost interest and moved away.

Although we rarely saw Chinki and Sanie together after this, we did on two other occasions see older females interact in a similar way with their younger half-sisters. In all three cases the younger animals disappeared from their clans a short while later. These few records are by no means conclusive evidence, but I have often wondered if this isn't the way that older females evict younger ones from their natal clan. If there is competition among the females for breeding opportunities, and obviously there has to be a limit on the number of females that can breed in a clan, then it might make sense for an older daughter to discourage her younger relatives from staying in the clan. As it is, it seems that the matriarch has the best breeding opportunities, as Normali produced five litters in seven years, whereas Chinki only managed two in five years.

The year 1977 was full of surprises. The next one occurred about a week after the altercation between Chinki and Sanie. We came to the den just after midnight to find Normali there with yet another strange male in close attendance. When she stood up to chew a bone, the male moved too and lay down next to her; when she moved to the other hole, the male followed. One of Normali's cubs came up to the male and presented its hind region as they do during the meeting ceremony. If we didn't know otherwise, we would have assumed that this male was a regular member of the clan. He stayed around the den for several hours before moving off.

Besides this male and the two that had mated with Normali the previous year, we had over the years built up a file of strange males that seemed to pass through the Kwang territory on an irregular basis. We even managed to catch one of these males and fit a radio collar to him, but we never saw him again. We called them nomadic males. By the end of the study we had identified 18 adult brown hyena males in the Kwang Clan's territory that appeared to be nomadic. Of these, 11 were seen once and never again, and 7 were seen again in the Kwang territory at least once within a month of the original sighting. Further sightings of three of these latter males were particularly interesting, as they were seen again in the Kwang territory 14, 18 and 27 months after the first sightings. Data from animals trapped and marked suggested that 8 per cent of the brown hyena population and 33 per cent of the adult males in the population were nomadic males.

In most species nomads are surplus animals that have been evicted from a territory and have little hope of breeding. In brown hyenas it seems as if these nomads are often responsible for mating with the group-living females. Besides the previously mentioned matings between Normali and two strange males, we saw another four mating bouts during the study, all of which involved strange males and known group-living females. Furthermore, there was a definite tendency for more strange males to be seen in a territory when a female was on heat than at other times.

This very unusual mating system seems to be a special adaptation for a species which lives at a low density and whose mating opportunities are few and far between. Because brown hyenas are solitary foragers, live in large territories and breed irregularly (never more than once a year and sometimes only once every four years) and non-seasonally, it may not be a good strategy for a male to invest time and energy in defending a territory and its female or females, as there is every chance of his being cuckolded. Rather he should become a 'travellin' man' and take his chances if and when the opportunity arises. As far as the females are concerned, they can either raise the young on their own or sometimes receive help from other family members, so they are not dependent on a mate. In times of food shortage, they might in fact be better off without one, as he could compete for food with the female and her cubs.

Early in 1988 we caught and marked Normali and Chinki's cubs. Normali's were given the names Hyaena, Brunnea and Thunberg, after the scientific name *Hyaena brunnea* for the brown hyena and the Swedish naturalist Thunberg, who first described the species. Hyena and Brunnea were females and Thunberg was a male. Chinki's three sons were named Lazarus, because we thought he died when we immobilised him; Bing, after the director of New Zealand Parks who had been with us when we caught him; and Elvis, after the King of Rock. At about this time Normali gave birth to a new litter of four cubs.

The new den was about a kilometre from the communal den at which Chinki's cubs were still residing for much of the time. We had collared Normali's son Shimi as I was interested to see if he too would carry food to cubs. He proved to be a dedicated helper. Not only that, but he gave us some insight into the question of helping and the degree of relatedness between the animals involved. Shimi was more closely related to Normali's cubs, with whom he shared 25 per cent of his genes, than he was to Chinki's, with whom he shared only 12.5 per cent. Kin selection theory would predict that Shimi should be more prepared to feed Normali's cubs than Chinki's.

We saw Shimi carry food to a den in his natal clan on four occasions. The first occasion was to the communal den. As he arrived, one of Chinki's cubs came running over to him and he merely dropped the food, which the cub ran off with. On the other occasions Normali's cubs were in the new den. On the second and third occasions Shimi came to the communal den at which only Chinki's cubs were living, but when the cubs came running over to him he was reluctant to give up the food. In fact he gave every impression of wanting to get past these cubs and go to Normali's ones. On both occasions he was unable to keep the food and was robbed by the cubs. On the fourth occasion he made an obvious wide detour around Chinki's den and took the food to his closer relatives.

It was at this latest den of Normali's that we also saw some interesting behaviour changes in sub-adults from being food 'providees' (being provisioned) to providers. The denning period for a brown hyena cub is in the region of 15 months. We regarded brown hyenas between 15 and 30 months as sub-adults. Sub-adults, although basically independent, spend quite a bit of time at the den. Should an adult arrive with food they will often rob the carrier and run off with it before the cubs can get a look in. One morning at sunrise Normali came to her den carrying a bat-eared fox. Chinki's 16-month-old son Bing was there. He grabbed the fox out of Normali's mouth and ran off with it before Normali could stop him. However, within minutes Bing returned with the fox. Normali snapped at him and he dropped the carcass into the den.

On another occasion, Hop-a-Long arrived at the den with a springhare. This time Thunberg, Normali's 22-month-old son from the previous litter, was at the den and came running over to Hop-a-Long emitting the harsh whine of a begging-call. Hop-a-Long dropped the springhare, Thunberg snapped it up and carried it away. As if his conscience troubled him, he returned half an hour later with the pelvic girdle and legs of a springbok, which judging by the amount of sand adhering to them had been stored in a hole close by. He came over to Hop-a-Long, dropped the food in front of him, presented to him submissively, then picked it up again and carried it over to a cub, before dropping the food in front of it. The cub picked up the 'second-class' meal and ran into the den with it.

In March 1978 the Kwang Clan stood at the all-time high of 16 members: the old male Hop-a-Long; Normali's two adult sons, Charlie and Shimi; the matriarch herself and four of her daughters, Chinki, Sanie, Hyaena and Brunnea; the sub-adult males Thunberg (Normali's), Elvis, Bing and Lazarus (Chinki's); and Normali's latest litter of four cubs. This had evolved from the original five – Normali, her three cubs and Hop-a-Long – in January 1974. In spite of the size of the clan, the territory that year, at 215 square kilometres, was smaller than it had been in previous years, when the membership of the clan was smaller.

As mentioned in Chapter 3, there seemed to be no pattern in either the size of brown hyena clans or the size of their territories. Our observations of the Kwang Clan confirmed this and also showed that this variation occurred both between different clans and within the same one. I had, of course been thinking about this for some time. Other carnivore researchers, notably Hans Kruuk with European badgers and his ex-student David Macdonald with red foxes, had found a similar phenomenon. Hans and David had, in fact, developed a hypothesis which they called the resource dispersion hypothesis, or RDH for short, to explain this. Briefly the hypothesis predicts that group size and territory size are regulated by the manner in which resources, particularly food, are distributed in that territory. If you think of food being available in patches, then, Hans and David said, the size of the group inhabiting a territory depends on the richness of the food patches, whereas the size of the territory depends on the distance between the patches.

Take, for example, a brown hyena food patch as either a few pieces of bone or a wildebeest carcass not killed or fed on by other carnivores. Clearly the latter is a far richer food patch than the former. During 1978 the Kalahari began to experience drought conditions after years of high rainfall in the early 1970s. Consequently many of the herbivores, in particular the wildebeest, whose numbers had risen during the times of plenty, began to die of starvation. This was particularly marked in the Kwang Clan's territory and left Normali and company with lots of nice big food patches. Consequently their numbers rose in this area. At the same time the average distance that brown hyenas had to walk between food patches decreased as more wildebeest carcasses became available along the Nossob river-bed, and with it so did their territory size. Cicely and Floppy-Ears from the Seven Pans Clan had a far larger territory than the Kwang animals, as the distance they moved between meals was far larger than that of their Kwang counterparts. Brown hyenas seem to fit the resource dispersion hypothesis, which goes a long way in explaining the fluctuations and differences in the size of the clans and their territories.

During 1978 we started to find several brown hyenas collecting at carcasses. The record number visiting a carcass on any one night was 12, something we had never seen in previous years. These of course were the rich food patches. Early one splendid, bright full-moon night, of the type that is unsurpassed in the Kalahari, we found Normali lying in the Nossob river-bed close to Kwang Windmill. It was unusual for a brown hyena to be lying down in the early part of the evening as this is the time they are most active. We stopped the vehicle close by and waited for her to become active. Within a few minutes we heard growling noises and immediately we knew that there were some lions feeding in the vicinity. We soon located three of the Kwang lion pride females with four large cubs, feeding on a fresh gemsbok carcass. This explained Normali's inactivity so early in the evening. We then noticed another two brown hyenas, Charlie and Chinki, lying on the opposite side of the carcass from Normali. It wasn't long before Shimi and Thunberg had also arrived on the scene. Typically they ignored the other brown hyenas there, even though they were close relatives and knew one other well, and each lay alone at distances varying from about 50 to 200 metres from the feeding lions. This was very different from what we had observed at the beginning of the study when a brown hyena, on finding lions feeding on a carcass, would immediately move away, returning periodically until the lions had finished eating and moved on.

After about an hour the lions suddenly moved away from the carcass when there was still a lot of meat left. We could hardly believe our eyes and, I should imagine, neither could the hyenas! Quickly but carefully they approached the carcass, until we had the unique sight of four of them tucking into a fresh gemsbok carcass. The party was, however, short-lived. The lions had merely gone to the windmill for a drink. After not more than five minutes of feasting, the hyenas scattered as one of the lionesses came bounding up to the carcass. She was soon joined by the others and the brown hyenas settled down to wait things out again. It wasn't until after sunrise that the lions finally abandoned the carcass, leaving precious little meat. By this time Normali and Chinki had given up waiting and had moved off separately to investigate other feeding opportunities.

Charlie was first onto the carcass. Brown hyenas have difficulty feeding on the remains of large carcasses, particularly in breaking off and consuming the larger bones. Charlie got to work on a back leg, obviously trying to dismember the leg from the pelvic girdle. We could not help thinking that his job would have been made much simpler if another hyena joined him. With two or more hyenas feeding on a carcass and tending to tug it in different directions, the carcass would be pulled apart with comparative ease. But Shimi and Thunberg appeared content to lie dozing in the early morning sun.

Eventually Charlie succeeded in detaching the leg. Immediately this happened, Shimi came over to the carcass and Charlie carried the leg off to eat it some distance away from the centre of activity. Shimi had been feeding for almost half an hour when Thunberg approached the carcass, circling downwind as he got closer. Shimi raised his hair and so did Thunberg a little. Thunberg stopped about three metres from the carcass and stood looking at Shimi, who continued feeding. Then Thunberg slowly came on to the carcass, so that the two hyenas were feeding at opposite ends, giving the distinct impression that neither enjoyed the company of the other. After four minutes Shimi moved away and lay down close by. It was the usual story. Brown hyenas obviously prefer to feed alone, in spite of their collecting together at a carcass. It is strange behaviour, perhaps bearing testimony to solitary origins.

In many social animals a dominance hierarchy is clear among the members of a group. This is particularly evident during feeding when the dominant animals have priority of access. As we were to learn and will discuss later, this is a striking feature of spotted hyena society but is not found in brown hyena society. Access to food seems to be on a first-come first-served basis and one animal will usually be content to wait for the other to finish eating before it feeds. There is not much point in wasting energy fighting over the food as there will be some left over when the incumbent feeder leaves.

During the latter half of 1978, five of the Kwang Clan members disappeared. As we might have expected, three of these were the young females Sanie, Hyaena and Brunnea, which probably dispersed in order to try to establish their own clans. Two others were cubs which mysteriously disappeared simultaneously. The final one was old Hop-a-Long.

TOP: Confident and aggressive, two spotted hyena females, their faces covered in blood from a recent kill, display their dominance.

BOTTOM: A brown hyena shows off its delicate hues in the early morning light.

TOP: A bright-eyed and bushy-headed Gus with the first brown hyena we caught.
BOTTOM LEFT: A trapped brown hyena. In 666 trap nights, we caught 29 different brown hyenas 169 times.
BOTTOM RIGHT: Houtop and Gus weighing an immobilised brown hyena.

TOP: Black-backed jackals often compete with brown hyenas when scavanging.
BOTTOM: A large cub with rabies attacks two adult females. This outbreak exterminated a small clan in the Auob riverbed. Rabies may be an important factor depressing spotted hyena numbers in the Kgalagadi Transfrontier Park.

TOP: Spotted hyena cubs are black at birth, the spots only begin to appear at about two months.

BOTTOM: For the first 12 months of life, mother's milk forms the bulk of the cubs' diet. Female spotted hyenas produce the protein-richest milk of any terrestrial carnivore.

TOP: Brown hyena cubs are exact replicas of the adults. Usually only a single litter is raised at a den in the Kalahari.

BOTTOM: Spotted hyena cubs grow up in a communal den with several litters of cubs, although a female will only have one or two cubs at a time.

TOP: The Battle of Kousaunt between two factions of females that saw the splitting up of the clan.
BOTTOM: A fight for dominance between two immigrant males in the Kousaunt Clan watched by the females.

TOP: Two adult spotted hyena females indulge in the bizarre meeting ceremony where they mutually sniff at each other's sexual organs.
BOTTOM: During a period of food abundance Normali and her daughter Chinki raised cubs in the same den.

TOP: Spotted hyenas often bravely stand up to their larger competitors, lions, and give as good as they get.

BOTTOM: Adult male lions, in particular, sometimes kill hyenas – especially cubs.

TOP: Four members of the brown hyena Kwang Clan share the remains of a gemsbok carcass – a rare sighting of brown hyenas feeding together. Note the radio collar and beta light on the second hyena from the left.

BOTTOM: A brown hyena pasting and a close up of the short-acting brown and long-lasting white substances secreted.

TOP: Two spotted hyenas seriously challenging an adult gemsbok – one of them is collared.

BOTTOM: A brown hyena carries the remains of a springbok killed by cheetahs back to the den to feed the cubs.

TOP: A rare occurrence of a brown and spotted hyena sharing a carcass. It didn't last long and surprisingly the spotted hyena soon moved off.

BOTTOM: When three kudu bulls, rare visitors, moved into the Kousaunt area, the spotted hyena clan killed two in four nights – testimony to the opportunistic nature of these wily predators.

TOP: Ostrich eggs are highly prized food items for brown hyenas. Here the old female Cicely hit the jackpot and found a nest with 26 eggs.

BOTTOM: Unlike brown hyenas, spotted hyenas are unable to bite open ostrich eggs unless they crack them first by kicking them together.

TOP: Gus's spotted hyena camp near Kousaunt.
BOTTOM: Margie and Michael visit Gus at his Kousaunt camp.

TOP & BOTTOM: The Kousaunt spotted hyenas became very habituated allowing us to occasionally enjoy some close-up encounters.

TOP: Margie and Gus adjusting a collar before fitting it to a hyena.
BOTTOM: Hidden holes, steep dunes and soft sand always held the possibility of the vehicle getting stuck while following the hyenas at night.

TOP: Margie, at 21, the happy camper in her new tented desert home.
BOTTOM: Gus with his two Kalahari boys.

We began to suspect that his time was nearly up when in July we watched him struggle for over half an hour to break open a springbok leg-bone, something a normal brown hyena will accomplish in less than five minutes. The reason was that his powerful bone-crushing teeth had become worn down with use and were becoming ineffective. Old hyenas get short in the tooth. Things obviously deteriorated for him and three months later he was dead. We found his remains at the den, suggesting that they had been carried there as food for cubs.

Over the years we recorded very few causes of death for brown hyenas. The most common cause was through injuries inflicted along the back, hind region and neck, as had been the case with the old Rooikop female, Ro-Ro. The only time we were certain who the culprits were was when the Kaspersdraai cub was killed by spotted hyenas. However, I am fairly certain that the injuries were mainly inflicted by other carnivores such as lions and spotted hyenas.

The mortality rate of brown hyena cubs is low: only 2 out of 16 we watched growing up did not survive to sub-adulthood. Sub-adult mortality is higher, most of the animals probably dying indirectly because they are unable to infiltrate a new clan or establish a territory. In the southern Kalahari a proportion of these are killed on surrounding farms by intolerant farmers. Once a brown hyena reaches adulthood, it has a good chance of living to old age, which seems to be around 15 years, when, as with Hop-a-Long, its teeth become useless.

In September 1978 our lives changed somewhat with the arrival of Michael, our first child. I now lost my field partner as Margie took up her new role of mother. At the end of the year I also started looking more seriously at spotted hyenas, although I was able to keep tabs on the Kwang Clan for another three years. I wanted to continue documenting the history of this clan in order to follow the fortunes of known individuals for as long as possible. I also hoped to be able to learn more about the elusive nomadic males. However, the most important issue regarding the brown hyena I had to deal with was to analyse and write up the results of the study. No work is complete until the results are published.

During 1979 the clan continued to lose members. Now it was the turn of the young males, as Shimi and Thunberg (Normali's cubs) and Bing and Lazarus (Chinki's) all disappeared. The bonanza of wildebeest carcasses was beginning to dry up and by the end of the year there were only six hyenas left in the clan: Normali and Chinki, Charlie and Elvis (Bing and Lazarus's litter mate), and the two surviving cubs: a male whom we had named Mistly (after our son *M*ichael *St*uart *Ly*ne), and a female Chrimoi (named after our friends *Chri*s and *Moi*ra who had been with us when we caught her).

Not having radio collars on brown hyenas any more made it difficult to contact our animals on a regular basis, so I had to make the most of any opportunities that came my way. I also occasionally went back to the original method of tracking spoor. Early in 1980 we tracked spoor to a new den in the Kwang territory. Once again it was Normali that had produced cubs, this time only two. By the middle of the year Mistly and Chrimoi had disappeared, leaving only four adults, Normali, Chinki, Charlie and Elvis, plus the two new cubs in the clan. In June I caught what I thought was another nomadic male in the Kwang territory and marked him for future recognition. By December this male, whom I called Thirdman (he was the third nomadic male I had marked that year), was still in the Kwang territory, which was most surprising. It began to look very much as if he had in fact joined the clan as an outsider, something we had never documented before.

Just after Christmas I found Thirdman near Kwang Pan. I had a dead springhare on the back of my truck, an unfortunate but fortunately rare road kill from the night before. If, I asked myself, I give this springhare to Thirdman, will he take it to Normali's cubs? The reason why I asked this question was that Thirdman was almost certainly not the father of Normali's cubs, as he had only appeared on the scene after they were born. After I dropped the springhare out of the truck three times, the hyena finally picked up its scent and came over to it. Predictably he started to eat the springhare immediately, but after a few minutes he picked up the remains and began walking off in the direction of the den. Normali's cubs were about 11 months old at this stage and when Thirdman arrived at the den there was nobody at home. He went down into the mouth of the den, dropped the springhare there and then moved across the mound and off into the darkness. My curiosity was satisfied and I went home.

As I drove home I thought about the significance of the behaviour I had just seen. If, as I have argued, altruistic behaviour, such as helping to feed cubs, has evolved through kin selection, why would an animal bother to feed cubs that were not related to him? Perhaps there was another type of mating system for brown hyenas apart from the nomadic male system we had unravelled. Perhaps Thirdman had chosen to immigrate to the Kwang Clan and to chance living with the two females, in the hope of being the first male to find them when either of them came into oestrus. The fact that Thirdman went on to spend at least four years living in the Kwang territory suggests that he may have gained the reproductive favours of the Kwang females.

This is one of the great fascinations about the study of animal behaviour. It does not matter how long you spend studying a species; sooner or later something happens to confound your ideas. Carnivores, with their highly intelligent and flexible behaviour patterns and social systems, are particularly interesting and rewarding study subjects. They are always coming up with new options to meet changing conditions.

At about the same time that Thirdman joined the Kwang Clan, quite by chance I rediscovered the Rooikop Clan. Early one morning I left Nossob Camp to do a routine game count along the Nossob river-bed. About 15 kilometres from the camp I came across a brown hyena carrying a large piece of meat. As I watched it climb a high dune and disappear over the top, I wondered if it was taking the food to a den. Later that day I returned to the spot and with the help of a tracker followed the spoor of the hyena. Not two kilometres from the river-bed we came to a brown hyena den. I could not help recalling the many days we had spent trying to find dens when the brown hyena study was in full swing, compared with the ease with which I found this one.

Like any carnivore I capitalised on my good fortune and, whenever I had the time, I visited the new Rooikop den. I was curious to find out which hyenas were using it. I was thrilled to discover another case of communal denning and even more so when I learnt that one of the two mothers was Phiri. Phiri was one of the two cubs at the old female

Ro-Ro's den whom I had marked nearly eight years previously and had not seen for nearly four years.

This was not the end of the exciting discoveries at this den. One evening while I was there an adult arrived with some food. I noticed that its ears were notched and carefully noted the position of the mark at each ear. I was a bit rusty about the identification of my marked animals but was pretty sure that it was Shimi, one of Normali's sons who had left the Kwang Clan two and a half years previously. When I got home I confirmed that the altruistic hyena was indeed Shimi. Here then was another example of a male joining a clan other than the one he was born into. It provided more evidence for immigrant males joining clans and for the alternative mating system I had previously advanced. Subsequently I gained evidence for immigrant males joining another two clans. Chinki's son Elvis joined a clan near Kaspersdraai and a young male called Elias, which I caught near Nossob Camp way back in 1974, turned out to have joined a clan near Groot Brak, 60 kilometres away from where I had originally caught him.

The highlight of our domestic life in 1980 came in September with the birth of our second son, Stevie. In February 1981 we left the Kalahari for a three-month visit to Scotland. I was in the final stages of writing up my doctoral thesis and the National Parks Board had kindly given me permission to spend some time at the Institute for Terrestrial Ecology at Banchory under the tutorship of Hans Kruuk. I also spent time at the University of Aberdeen's Culterty Field Station thanks to the hospitality of my good friend and colleague Martyn Gorman. I had met Martyn through Hans and we had been collaborating on a study of the scent-marking behaviour of hyenas. These three months were most stimulating and gave me a good chance to expose my data and ideas to other scientists, something which I was not able to do much back home. It was also a fine experience for the family to taste something of a very different type of life. My thesis was eventually handed in in December 1981. So ended a career of 17 years during which I was registered as a student at either Cape Town or Pretoria university.

The year 1981 was the last year that I was able to monitor the composition of the Kwang Clan in any detail. By the end of that year Normali and Chinki were still going strong. Normali was at least 10 years old and Chinki was nearly 8. Chinki had at last given birth to three cubs in late 1980, her first litter in three years, but I was unable to find the den in 1981 and don't know what happened to them. Charlie was also still around, as was the immigrant male Thirdman. I was also surprised to find Chrimoi, one of Normali's daughters, back in the territory after an absence of over a year. Had she been nomadic for a time before returning to her natal clan?

In January 1982 Charlie disappeared from the Kwang Clan at nearly eight years of age. I was fairly sure that he must have died, as he was far older than the normal age of 20 to 30 months at which most brown hyena males become nomadic. Once again I was proved wrong. At this advanced age he did indeed become nomadic, making brief appearances back in the Kwang territory 17 and 31 months after leaving.

When we finally left the Kalahari in May 1984, Chinki was still alive and over 10 years of age. I last saw Normali in September 1982 when she was at least 12 years old, but I think she must have died soon after that. We made a brief visit to the Kalahari in December 1984 and were fortunate one night to find Chinki foraging along the Nossob river-bed just north of Cubitje Quap. That was the last record I have of her.

CHAPTER 7

Getting to Know the Spotted Hyenas of the Kousaunt Clan

IF ANYONE HAD ASKED ME what species I wanted to focus on after the brown hyena study, I would not have hesitated to say the spotted hyena. Fortunately for me, this low density species in the Kalahari was a priority. Not only was it important to investigate the reasons for the low numbers of spotted hyenas in the area, but we also needed to study the role they filled in the Kalahari ecosystem and their importance as predators. From the few tantalising glimpses we had obtained into their lifestyle in the Kalahari, and from what Hans Kruuk had revealed in East Africa, I knew that they must offer a treasure trove of exciting facts. It would also be interesting to contrast their behaviour with that of the brown hyena in the same ecosystem.

Accordingly, in 1979 my priorities as far as hyenas were concerned switched from the brown hyena to their much maligned, larger spotted cousins. I decided that, as with the brown hyena study, I needed a main study clan on which to concentrate, as well as several other, less intensive, back-up groups on which to test certain ideas. In particular I wanted to test whether the resource dispersion hypothesis could explain the low density of spotted hyenas in the Kalahari.

I had to look further afield to find a suitable spotted hyena clan than I did for a brown hyena clan. The nearest was the Kaspersdraai Clan about 30 kilometres to the south of Nossob Camp, but lately these animals had not been in evidence. However, some 50 kilometres north of the camp around Kousaunt Windmill, there appeared to be quite a large clan. After making a few preliminary observations of these animals early in 1979, during which I established that they were relaxed in the presence of my vehicle, I decided that they were the best clan to concentrate on. A striking member of this clan was an old female whom I had known for many years. Large, one could say almost obese, with practically no hair on her body, both her ears tattered and torn, and her mouth hanging open so that she drooled permanently, she was probably the ugliest hyena I had ever seen – the original caricature of an unloved species. Little did I know how much I would come to admire this animal, the matriarch of the clan, over the next few years. I called her Old Flat Ear.

Because of the distance from home to Kousaunt I decided to build a camp in the area, so that I did not waste valuable kilometres driving to and from my study animals each day. Budget constraints seriously curtailed the distance I could drive each month. I also needed to engage the services of an assistant as Margie was, unfortunately, unavailable. I needed someone to accompany me on night-long vigils, as well as someone adept at tracking spoor. After one or two failures I acquired the services of a young man called Hermanus Jaggers. Hermanus was to prove an invaluable colleague who grew to become as fond of and captivated by spotted hyenas as I was. Although he was not a San, he turned out to be a fine tracker. Tracking spoor was to prove to be far more valuable in the spotted hyena study than it had been with the brown hyenas.

Once I had decided on my study clan I needed to mark the members in order to keep tabs on the fortunes of each. I was soon presented with a good opportunity to do this when tourists reported that they had found a number of spotted hyenas feeding on a wildebeest carcass at Groot Brak Windmill, the next one north from Kousaunt. Kapok, one of the game scouts from Nossob, and I arrived at sunset to find six hyenas, including Old Flat Ear, collected around the small drinking reservoir, next to which the carcass was lying, and one rather sorry-looking male hyena standing in the shallow reservoir. His tail was tucked between his legs and his ears were flattened. The hyenas standing around the reservoir looked far more confident, with their tails up and ears cocked, and were clearly focused on the unfortunate one in the water. One of them, a fine-looking male specimen, seemed to be particularly interested in him. Suddenly, he jumped into the reservoir and attacked the other, biting at his neck. This excited the onlookers, who started making a curious kind of grunt-laugh. The victim gave vent to a loud roar-growl as he tried to parry the advances of the attacker. After not more than 15 seconds the dominant animal left him and climbed out of the reservoir. The defeated animal remained in the water, looking just as miserable and bleeding slightly from the neck. During the next hour the attacking male made four more similar attacks on this male, which all the while remained in the pool.

As interesting as this behaviour was, we had other priorities that night – to examine and mark as many of the animals as we could. We off-loaded our paraphernalia at a tree some 50 metres from the hyenas and set about the night's work, hoping to be lucky enough to immobilise three or four. I had only a dart pistol with me, which meant that I needed to be within metres of my target in order to be reasonably sure of hitting it. We drove slowly towards the carcass on which three animals were now feeding, Old Flat Ear and two equally large but obviously younger females. The miserable male was still standing in the reservoir and the others were lying about in the vicinity. Closer and closer we inched without their paying us any attention. When I was close enough to be confident of hitting one, I fired. The three animals scattered as the dart struck home in the rump of one. She ran off about 10 metres before stopping. We moved away from the carcass and the other two immediately returned to it and resumed eating, while the darted animal walked off slowly. We followed her at a respectful distance. Seven minutes after I darted the hyena she started swaying gently and, after a few uncoordinated steps, she gently sank to the ground. We waited a few more minutes, then drove closer. We loaded her on to the back of the vehicle and drove over to the working tree.

She weighed 72.5 kilograms, nearly twice the size of a brown hyena, which weighs about 40 kilograms. We took other body measurements, looked at her teeth in order to get an idea of her age and cut a small nick out of one ear for future recognition. We then moved her to another tree close by to allow her to recover from the drug, which would have taken about an hour.

I wondered how the rest of the clan would now take to the approach of the vehicle. As the area we were working in was quite open, they had obviously seen us pick up the first hyena, transport her to the tree and work on her there. Surprisingly I was allowed to approach as close as previously. Once again I easily darted another large female, with more or less the same result as before. This was to be the pattern throughout the night. We managed to dart and mark eight hyenas without any trouble. The others paid very little attention to what was going on, if anything they were rather curious, and at one time approached to within a few metres from us at the working tree.

Of the eight animals we caught, three were adult females and the rest were young males, two years of age and less. I called the three females Ella, Olivia and Tu-Tu, Ella after Ella Fitzgerald, Olivia after Olivia Newton-John and Tu-Tu because her marking code was 2/2 and I also imagined her as a good ballet dancer. The males were named after people who had previously published work on spotted hyenas: Fritz after Fritz Eloff, who first pointed out the fact that spotted hyenas in the Kalahari were efficient hunters; Simon after Simon Bearder, who had made a study of spotted hyenas in Timbavati; Harrison-Matthews after the first person to properly describe the peculiar anatomy of the female spotted hyena's reproductive organs; and James Bruce after an 18th-century naturalist who wrote: 'There are few animals, whose history has passed under the consideration of naturalists, that have given occasion to so much confusion and equivocation as the Hyaena has done. It began very early among the ancients, and the moderns have fully contributed their share.'

The three remaining hyenas there that night were Old Flat Ear, who was easy to recognise without having to be marked, and the two adult males. We had become so focused on our capture operation that we did not notice the departure of the defeated hyena at some time during the night. I was not too concerned about him as he had a large chunk of ear missing, so I was confident of being able to recognise him. I was not sure that I would be able to recognise the second male that had also disappeared, perhaps to continue his aggressive and dominant behaviour towards the other. The significance of this aggressive behaviour between the two males would only become obvious to me some time later.

Over the next few weeks I learnt more about the clan. I soon found their den in the Nossob river-bed close to Groot Brak. Three of the females had cubs at the den. Olivia had two cubs of eight months old: a female called Goldie, after Goldie Horn, and a male called Silver because his sister was Goldie. Old Flat Ear had a son of about one year called Butch, after Butch Smuts, and Tu-Tu had a male cub of about three months old called Paka, after our friends *Pa*ul and *Ka*tinka. The only other animal in the clan that I had not marked was a male whom I caught and marked a week later and who from photographs I had taken on the epic night at Groot Brak proved to be the aggressive male. I called him Hans after Hans Kruuk. The defeated male was nowhere to be found.

After my experiences with brown hyenas I was sure that the only possible way to follow spotted hyenas at night would be with the use of radio collars. However, after fitting two collars I soon decided that radios were not really so useful in these conditions. On moonlit nights I was able to keep up with the hyenas without having to rely on radios, as there were often three or more animals in a foraging group. Moreover, their very large ranges made it difficult to locate radio-collared animals from the ground. The range of the transmitters was not more than five kilometres at best, and very often in the undulating dunes it was less than two kilometres. I discovered that the best way to find hyenas was to go to the den at sunset. Because the social denning habits of the spotted hyena are more marked compared with those of the brown hyena, there were usually some hyenas at the den and I would merely wait for them to move off.

My routine was to spend about eight nights a month around full moon following hyenas, while the rest of the month was taken up with my other duties as a research officer. I erected a fairly permanent shelter near Kousaunt, which became my day-time resting site. The experience I had gained from working at night with the brown hyenas, or rather the experience I had gained from sleeping in the day, stood me in good stead. There was no way that I could work for eight nights continuously if I was not able to get a good sleep during the day. In the winter this was not too difficult, but in summer when the temperature often rose to above 35°C in the shade, it took some getting used to. Lying on a wet towel provided really good air-conditioning. More importantly, I learnt not to get up if I woke after two to three hours of sleep. This was simply not enough. I just kept lying down with my eyes closed and almost forced myself back to sleep.

On some occasions my day-time slumbers were disturbed by bees. The bees were attracted to the water we brought to the camp. At first one or two would discover the moisture and then go back to the hive to call the others. For about 30 minutes there would be a swarm of bees in the shelter. The only way to deal with this was to keep my head down and my eyes closed, and I never got stung! Hermanus, too, became pretty adept at sleeping under these conditions, though most of my visitors suffered.

It was some time before I made my first observation of a full hunt, although I soon obtained some tantalising clues that they were indeed hunting frequently. On one of the first nights out we found Tu-Tu just after dark lying on the short bushman grass plain near Kousaunt Windmill. One big difference between brown and spotted hyenas, I was soon to learn, is that spotteds spend a lot more time sleeping than their smaller counterparts. Spotted hyenas only devoted 55 per cent of the hours of darkness to active pursuits such as socialising, hunting and eating, whereas brown hyenas were usually active for 80 per cent or more of the night. Watching spotted hyenas sleep was going to take up a fair bit of my time during the next few years. During these hours of inactivity the BBC World Service became a good companion, although on particularly inactive nights even the world news became rather boring when heard for the sixth time.

Notwithstanding the greater amount of leisure time spotted hyenas enjoy, they can change from complete relaxation to frenzied action in the blink of an eye. On one occasion, after lying motionless for almost three hours, Tu-Tu suddenly jumped up and ran off into the night. Luckily I was watching her at that instant. I quickly started the car and took off in the direction she had taken. Scanning with my hand-held spotlight as we raced across the plain, we were fortunate to find her again quickly. With her ears cocked, running at a steady 30 kilometres an hour, she had clearly heard something exciting happening. After we had gone nearly a kilometre, I saw the eyes of several animals shining in my car lights. As I approached closer I saw that they were the eyes of other hyenas and that they were circled round the corpse of a red hartebeest. The hyenas had blood on their faces, but the carcass did not appear to be host to any other carnivorous animals. Had the hyenas killed the hartebeest, or had it succumbed to some other agent of mortality? Judging by the full bellies of the hyenas at the carcass and the amount of meat still available, it was clear that the hyenas were the only animals to have fed on the carcass. The next morning I was able to confirm from looking at the tracks around the carcass that this animal had indeed been killed by the hyenas.

Several nights later we were following five hyenas in the dunes close to the Nossob River. Suddenly, almost by chance it seemed, they were in among a small herd of about a dozen gemsbok. There were quite a few large blackthorn bushes in this area and it was difficult to keep up with the hyenas. Gemsbok and hyenas were running all round the vehicle and my desperate attempts to keep some hyenas in sight were futile. I did not want to shine the spotlight too widely for fear of dazzling the gemsbok and giving an unfair advantage to the hyenas. Within seconds I had lost contact with the action. Fortunately one of the animals was wearing a radio collar and so I was able to change over to radio tracking.

The hyenas must have moved a good distance because I could only hear a faint signal. This made direction location difficult and, to my frustration, I quickly lost the signal completely. I was forced to drive up to the top of a dune nearby before being able to pick up a signal and get a direction on the collar. I eventually found the hyenas half an hour after the hunt had begun. They had killed a large gemsbok calf of about a year old. This was one of the few occasions when the radios were of help.

My observation period over, I returned to Nossob Camp, still waiting to experience a complete hunt. During the next full-moon period the hyenas were rather inactive and I spent seven nights with them without documenting a kill. I did, however, learn quite a bit about the social system of the Kousaunt Clan and the denning behaviour of spotted hyenas.

In contrast to the brown hyena, communal denning is the rule for spotted hyenas. The three mothers spent a lot of time at the den. The fourth female, Ella, was absent. Old Flat Ear and Olivia were particularly close friends. I am pretty sure that Olivia was Old Flat Ear's daughter. They usually lay together and were dominant over Tu-Tu, who always gave way to them. The young males were quite well tolerated around the den and spent time play-wrestling with the cubs. The large male, Hans, on the other hand, kept his distance and was subservient to the three adult females.

The females spent long periods nursing their cubs – sometimes a cub stayed for over an hour attached to a nipple. Unlike the brown hyena mothers that were not averse to suckling each other's cubs, the spotted hyena mothers were fastidious about suckling only their own cubs. Furthermore, it was striking that adults arriving at the den never brought food for the cubs. Sometimes, if they made a kill close by, someone might drag the remains back, but the skin and bones that these carcasses comprised were not suitable food for the cubs. Neither did the cubs go foraging on their own or with the other members of their clan. I was to learn that spotted hyena cubs rely almost entirely on their mother's milk until they are nearly one year old. This is a strong trend in spotted hyena society that has since been reported from several other studies. The only instance I am aware of where this rule has been broken happened in the Kousaunt Clan several years later, after we had left the Kalahari.

My successor as research officer in the Kalahari, Mike Knight, kept tabs on some of the hyenas. By this time, as we will see later, the Kousaunt Clan had undergone some drastic changes. Of the original females only Olivia remained. The other adult females of the clan were her daughters and granddaughters. At this time three of them had one cub each. Times were hard as prolonged drought had led to a decline in prey and resulted in relatively high cub mortality. For the only time in recorded spotted hyena history, two of the three females were seen several times to suckle cubs that were not their own. By now I am sure I do not have to mention that this behaviour can be explained in terms of kin selection theory. During this stressful period, females that were close relatives and had only one cub each were prepared, when they had spare milk, to give it to a close relative's offspring.

But back to 1979 and the observation period I have been talking about. On the last night out I found several hyenas feeding on a scavenged springbok carcass. I was pleased to see the missing female, Ella, with them and interested to note that her teats were swollen. She did not stay long and I followed her when she moved off. The carcass was in the river-bed, but she immediately moved into the dunes. It was quite a business following her as she moved quickly, loping along at a speed of 10 kilometres an hour. Fortunately she moved in a straight line and I was able to maintain visual contact. After 5.5 kilometres she came to a hole and called out two very small cubs. This was her nursery den, away from the hustle and bustle of the communal den. When I started observations in the next full-moon period Ella had moved her cubs in with the rest of the gang.

This was also the first observation period that Hermanus worked with me. On the first night out we were fortunate to find five hyenas at Kousaunt Windmill before dark. They obviously meant business as they soon moved off. They travelled south down the Nossob river-bed and over an open plain in single file. The five dark silhouettes strung out in a line against the pale grass in the moonlight was a beautiful and evocative sight.

They continued in a southerly direction for nearly two hours without once stopping to rest. Suddenly they changed direction and started to lope, their heads up, noses in the air, obviously sniffing something. There was a slight breeze blowing from the east and this was the direction they took. They were now running at about 20 kilometres an hour, ears pricked forward and eyes focused in front of them, bushy black tails curled over their backs – a sure sign that something had excited them.

They kept this up for nearly three kilometres and then, as we came over a small dune, a herd of about 10 gemsbok appeared in front of us. The hyenas had apparently smelt them from afar. The hyenas accelerated, as did my pulse rate, and the gemsbok scattered. Things were now moving so quickly that it was all I could do to keep behind one of the hyenas. In the parking lights and moonlight I could make out another two about 20 metres ahead, hard on the heels of a gemsbok. I glanced at the speedometer, which showed that we were travelling at 40 kilometres an hour across the open veld.

Though the gemsbok appeared to be pulling away from the hyenas, the latter showed no sign of giving up. After a while the gemsbok turned slightly to the right. The hyenas immediately picked this up and adjusted their direction so that they cut the corner and began to make up some of the difference between them and the gemsbok. The chase had now gone on for 2.5 kilometres and the gemsbok appeared to tire as the hyenas rapidly gained on it. As the three hyenas caught up with the gemsbok it tried to back up against a dead tree and face them. However, Hans grabbed it in the groin, at which point Olivia rushed in and knocked the antelope down onto its side. The third hyena, one of the young males, remained, like us, a spectator.

It was a young gemsbok of about one year old, weighing close on 100 kilograms and with horns about 80 centimetres long. The gemsbok tried to butt the hyenas with its horns but, being on its side, had difficulty in using them effectively. Soon the hyenas had pulled out the gemsbok's stomach and it was dead. The other two hyenas came running in from the dark and the feast began. Bolting down chunks of meat, their faces quickly covered in a macabre mask of blood, the hyenas polished off the best part of the gemsbok in an amazingly short time. After two and a half hours, all that was left was the head, some skin and the larger bones. Anyone arriving at the scene at this time would probably have assumed that lions had vacated the carcass and left the remains to the scavengers.

It had been an incredible few hours. The actual kill was only a small part of the entire incident. Taken in perspective, it was a fitting climax to a number of events: hours of searching for the prey, detecting it, cutting it off from the rest of the herd, and a long, hard chase. And of course the hyenas had done it for one reason only – to obtain food in order to survive. Over the next few years I was to witness many similar incidents, and each filled me with excitement and admiration for these superbly tuned hunters.

What about the victims? Death is the only certainty in life for man and beast, and being killed by a predator is probably one of the quickest ways to die, certainly preferable to dying of thirst or starvation.

CHAPTER 8

Hunting and Other Foraging Fables about Spotted Hyenas

HERMANUS AND I DOCUMENTED nearly 150 kills by spotted hyenas – not bad for an animal that is still popularly regarded as a scavenger. Many of these were witnessed directly, as I have described in the last chapter. Others were discovered when we followed hyenas to a kill made by other members of a clan. A third method was by tracking spoor.

We had far more success tracking spotted hyena spoor than we had had with the brown hyenas. Spotted hyenas do not meander, nor do they investigate bushes, holes and other sources of small pieces of carrion as brown hyenas do. They normally move straight, only making large-scale direction changes when they detect prey. Also, they very often move in a group and the tracks of several animals moving together are obviously far easier to follow than the tracks of a single animal. We were thus able to collect a lot of information about the hunting behaviour and movement patterns of the spotted hyena by tracking spoor.

Like most children who grow up in the Kalahari, Hermanus had spent much of his childhood minding his father's sheep and goats. If any animals strayed from the herd the children would find them by tracking their spoor. As I have said, Hermanus is a very good tracker with great powers of endurance. His eye was so quick and focused on following a spoor that for much of the time he could sit on a special seat that Rian Labuschagne, the ranger at Nossob during the latter years of the study, constructed in front of the bonnet of my Land

Cruiser. By making hand signals Hermanus would show me where to drive, and by the position of the sun I was able to record the direction of movement for Margie to plot on a map – not as accurate as a GPS but still quite effective. Much of the information we gained from clans other than the Kousaunt one was through tracking spoor.

Spotted hyenas eat almost any kind of animal matter and even sometimes indigestible items such as paper and motor-car tyres. Their habit of scavenging from rubbish dumps and around towns is legendary; they have even been recorded digging up human corpses. Traditionally many African communities left corpses out in the bush for spotted hyenas to dispose of. From time to time people sleeping under the stars have been attacked by spotted hyenas, but these are comparatively rare events and I know of only one case where a hyena ripped through a flyscreen and dragged a person out of a tent. There is one notorious case of spotted hyenas becoming habitual man-eaters. This occurred in the Mlanje region of Malawi during the late 1950s when over a seven-year period five to eight people a year were reportedly killed by hyenas. These attacks took place during the summer months when people were sleeping outside their huts.

In spite of its reputation for having such a catholic diet, the spotted hyena is a rather specialised feeder in most areas, far more so than the brown hyena. Gemsbok and wildebeest and, to a lesser extent, springbok, hartebeest and eland make up the bulk of its food in the Kalahari, and of course, contrary to popular opinion, most of these animals are killed by the hyenas, not scavenged. I calculated that 73 per cent of the Kalahari spotted hyenas' food came from their own kills.

Neither hyenas nor any other predator can simply go out and kill the first prey animal they find. The prey have evolved a number of strategies and behaviours for escaping predation, and the predators are sometimes hard put to break through its defences. The balance between predator and prey is a fine one, with neither having the upper hand. The evolutionary arms race, as it has been called, is a most fascinating aspect of animal behaviour.

Our hyenas hunted mainly gemsbok and wildebeest, although eland calves, when they were present in an area, attracted much attention. Springbok, except for lambs, and hartebeest were rarely hunted. They are simply too fast for the hyenas. Most of the members of these two species eaten were scavenged – remains from the kills of other predators or animals that had died from starvation and disease.

Young animals are, not surprisingly, particularly vulnerable to the jaws of hyenas: 70 per cent of all the kills we witnessed were juveniles younger than one year. But they are also capable of bringing down adult gemsbok, wildebeest and, even on occasions, eland. Gemsbok calves top the Kalahari spotted hyena's menu, making up 43 per cent of the kills we witnessed. Their success rate at catching them was high: 63 per cent. This is somewhat higher than the success rate usually achieved by lions.

When spotted hyenas encounter gemsbok with calves, the gemsbok immediately close ranks with the calves in the middle. They stand up to the hyenas, parrying and thrusting at them with their rapier-like horns as the hyenas attempt to break up the formation and cut a calf out of the herd. This rarely happens. However, if, as hyenas approach, the herd scatters and the calves are unable to close formation with the adults, they have to run for their lives. In only 9 out of 49 times that we saw this happen did a calf manage to outrun a hyena. Although they may be faster than the hyenas over the first kilometre, the superior stamina of the hyena almost always wins the day. The average chase in a successful hunt of a gemsbok calf was 1.7 kilometres, and the older the calf the longer the hyenas had to run for their supper.

The crucial factor in a gemsbok calf hunt is thus whether the hyenas can separate a calf from the adults. One might think that to this end the number of hyenas hunting and the number of gemsbok in the herd are important. I was surprised to find that neither was the case. A single hyena was just as likely to catch a calf in a herd of 20 gemsbok, as the five hyenas hunting together were to select a calf out of a herd of five gemsbok. It appears that the most important factor determining the success of a gemsbok calf hunt is the manner in which the antelope are distributed when the hyenas first arrive: the more scattered they are, the better the hyenas' chances.

If hyenas are able to separate more than one calf from a herd, they may chase the calves independently of each other. This sometimes led to the hyenas bringing down two calves. Once, for example, Hermanus and I tracked the spoor of two hyenas into a herd of about 10 gemsbok. We could clearly see that each hyena had selected a calf. The first hyena pulled down its calf after only 450 metres, but the second one had had to run over two kilometres for its calf.

Gemsbok herds with calves are not always encountered by spotted hyenas. More typically solitary bulls, or two or three adults of either sex, are met up with. The encounters are usually brief: the hyenas approach, the gemsbok stand their ground facing the hyenas, and the hyenas soon move away. Sometimes the gemsbok run away from the hyenas. When this happens the hyenas run after them. Again these chases often fizzle out, and after one kilometre the hyenas have either given up or the gemsbok have stood at bay.

However, if the chase goes on for longer than one kilometre, things become more serious and the outcome of the hunt becomes less predictable. Both parties run faster and, if the gemsbok stands, the hyenas try harder to overpower it. Still, we never saw hyenas kill a gemsbok that stood at bay.

When facing hyenas, the gemsbok often backs up against a tree to protect its hind region. Sometimes they will even go right into a circular clump of candle acacia thorn bushes, making it very difficult for the hyenas to get at them. Then it becomes a waiting game for the hyenas to run out of patience or the gemsbok to run out of nerve. Usually the gemsbok wins. Once I watched hyenas waiting over five hours for the gemsbok to come out, but to no avail.

On the only occasion that we saw a gemsbok leave a thorn bush clump, the hyenas immediately went after it again. After about one kilometre the gemsbok took refuge in another thorn clump. However, about 30 metres short of the bush one of the hyenas grabbed hold of its tail. The hyena bit too hard and, lizard-style, the gemsbok escaped tailless into the bush, leaving a snack for two of the hyenas. The gemsbok learnt her lesson and did not budge until the hyenas finally moved off one hour later.

The only adult gemsbok we saw spotted hyenas kill were those that could not outrun the hyenas nor did they stand to face them. This happened in only 7.6 per cent of the encounters. All these gemsbok were killed relatively easily. The hyenas seemed to pull the gemsbok down on to their sides, so that their horns became ineffective.

Why should some adult gemsbok face up to spotted hyenas with impunity, while others run first then face them, and still others are killed relatively easily? I think the answer lies in the condition of the gemsbok. A healthy and fit adult gemsbok can confidently stand up to hyenas – it knows it is strong enough to defend itself and could even kill a hyena with its sabre-like horns. Even if the gemsbok is slightly vulnerable it can defend itself, albeit with a bit of a struggle. Perhaps, those gemsbok that are unfit, unwell or old do not have the confidence to face hyenas, and so keep running. Because they are in poor condition the hyenas soon catch up with them and before they can get into a position to defend themselves the hyenas bring them down. Endowed with fine eyesight and able to determine quickly the condition of the prey, the hyenas are able to size up the situation and react accordingly.

We did collect some evidence to support this idea. Firstly, the distances that adult gemsbok were chased when they were killed were quite short, on average 1.6 kilometres. Secondly, by examining the marrow in the bones of hyena-killed gemsbok we found that they had low fat reserves, which is indicative of an animal in poor condition. Animals store fat in a number of places, around the stomach and kidneys and in bone marrow. The fat in bone marrow is the last of the fat reserves to be drawn on, and once this is gone the animal is on the way to starvation.

After gemsbok, wildebeest are second on the spotted hyenas' hit-list. Although they kill mainly wildebeest calves, adult wildebeest seem easier for spotted hyenas to overpower than adult gemsbok. The distance that hyenas had to run before pulling down wildebeest was usually only about one kilometre. Unlike gemsbok that stand at bay, the best chance a wildebeest has of escaping from hyenas, once it has been singled out of the herd, is to try to outrun them.

The first hyena kill I ever saw was a wildebeest bull, long before I started the spotted hyena project. I was driving up the Nossob river-bed from Twee Rivieren one night when I came across six hyenas around a wildebeest bull standing in a rain puddle in the middle of the road. At first I could not make out what was happening until I noticed that the wildebeest was bleeding from its anus. Then one of the hyenas darted under the wildebeest's belly and bit it in the stomach. One or two of the others joined in, in a rather half-hearted manner, while the rest just stood by and watched. The wildebeest did not appear to be putting up much resistance and merely remained standing. After a minute or two it sagged on to its belly, but the hyenas seemed unable to attack it while lying in this position.

However, by harrying the wildebeest, running around it and darting in to bite its flanks and tail, they managed to get it to stand again. Then one of the hyenas was caught by a swinging horn and tossed about a metre into the air. This did not seem to deter it and it came right back. By now some of the other hyenas were making a more determined effort to kill the wildebeest. Its abdomen had been ripped open and its innards were hanging out, yet it still remained standing. Eventually, 25 minutes after I arrived on the scene, through exhaustion and loss of blood the victim dropped down on to its side and was devoured by the hungry hyenas. I was rather surprised that the hyenas had struggled so long to kill the wildebeest, which had really done little to ward off its attackers. Most of the other wildebeest kills I observed were far quicker and cleaner than this one.

One of the most incredible feats of strength I saw a spotted hyena perform was by the female Ella. She and her two one-year-old cubs chased a wildebeest bull. They caught up with it quite quickly and Ella, almost single-mouthed, killed it. Her cubs merely ran around watching the struggle. When she had finally killed the animal, after about 20 minutes, she was exhausted. She lay on her side panting heavily for 10 minutes, even kicking sand under her belly to help her cool down.

The spotted hyenas paid little attention to the large numbers of springbok they often shared the river-bed with. Adult springbok are, quite simply, too fast for hyenas. It was only during the lambing season that hyenas really became serious about hunting springbok. Then their hunting behaviour resembled that of the brown hyena when they were after lambs, except that spotted hyenas were successful in 31 per cent of the hunts, whereas brown hyenas were only successful in 6 per cent.

The most nomadic spotted hyena prey species in the Kalahari is the eland. For months we would not see a single eland, then out of nowhere they would appear, sometimes in their thousands. After a few weeks they would just as quickly disappear. Unlike the movements of springbok and hartebeest, which come into the river-beds during the rains and move away during the dry season, eland movements are unpredictable. Whenever they came into our study area the hyenas paid them a lot of attention. In terms of the meat they provide, both eland adults and calves are profitable hunting fare for hyenas. Eland calves seemed to be more rewarding for hyenas to kill than gemsbok calves in this regard, and I think that the hyenas hunt them in preference to gemsbok.

Because of their large size, adult eland are difficult for hyenas to overcome. The few times that we watched hyenas and adult eland interact were dramatic, and I was interested that the hyenas paid such attention to these massive beasts. The first time I saw this I was following Old Flat Ear and two young males. They encountered a young eland bull and chased it a few hundred metres when he stood and faced them. The hyenas darted in and out, trying to bite the eland at the shoulder and the rear as he swung his horns at them. Even though they managed to score a few blows, they did not appear to be making any impression on the eland.

After a while it ran off again and the hyenas chased it, now trying to hamstring it. Again the eland stopped and defended itself by swinging its horns and kicking out vigorously with its back legs. One connection of a hoof to a hyena's head would surely have ripped it open. For the next two hours the hyenas continued to harass the eland, sometimes vigorously and at other times less so. Whenever the eland moved they followed – one hyena at the front and two behind – just waiting, it seemed, for the eland to put a foot wrong. Eventually after moving 11 kilometres with the eland and with the sun rising, the hyenas gave up.

Others have witnessed similar seemingly futile hunting attempts by spotted hyenas on very large prey. Hans Kruuk saw hyenas in the Ngorongoro Crater attack black rhinos and their calves and Joh Henschel saw them attack giraffe and hippos in the Kruger National Park. One of Joh's observations shows why they do it. Four hyenas encountered a hippo and followed it as it ran off in what appeared to be a futile exercise. Suddenly it stumbled and slipped badly, seemingly to dislocate its shoulder. That was all the hyenas needed and they rushed in to finish off the unfortunate hippo. As a result of this opportunism the hyenas were provided with a feast for the next week. No wonder they are Africa's most successful large carnivore. However, when you play for such big "steaks", you take a high risk. A friend, Tony Starfield, told me that he once watched a hippo bite an attacking hyena's head off in Savuti, Botswana.

The supreme example of the tenacity and persistence of spotted hyenas was shown to Hermanus and me when we tracked the spoor of three hyenas that had followed a herd of about 20 eland for almost 25 kilometres. Eventually, after challenging the eland several times, probably in a similar manner to what I observed with Old Flat Ear and the two young males, they got hold of a young bull. In a titanic struggle, which took place over a distance of a kilometre, they eventually killed it.

Another example of the peerless opportunism of these fascinating creatures occurred when three fine kudu bulls visited the Kousaunt Clan's territory. Kudu are rare in the southern Kalahari, although their numbers increase to the east in Botswana. The hyenas soon discovered these rather strange-looking new arrivals and in the space of four nights killed two of them.

From time to time spotted hyenas hunt prey that they usually ignore. Bat-eared foxes are one example. On five occasions we saw solitary hyenas chase these small canids. Three times the foxes easily outran the hyenas or scurried down a hole. Once a hyena chased a fox for about 500 metres, during which the fox stopped abruptly three times to face the hyena, with its tail raised in an inverted U and uttering a rattling growl, in a manner similar to what I described earlier when brown hyenas hunt them. Each time the hyena stopped and appeared to get a fright. The fifth chase after a bat-eared fox had a most unfortunate ending. The fox suddenly stopped and doubled back past the hyena, only to run straight into my vehicle. The hyena ran up to the stunned fox and with one shake killed it. It then lay down next to its victim without so much as taking a nibble. After an hour a second hyena arrived. It sniffed at the fox, rolled on it and then moved off with the original hyena. Unlike brown hyenas, spotteds do not seem to find canids particularly appetising.

Remembering the brown hyena cub Bop's successful striped polecat kill, I once saw two sub-adult spotted hyenas encounter one. They chased it for 200 metres. Several times one of the hyenas made as if to bite it but drew back quickly. Once one of them even picked up the polecat and tossed it into the air a metre or two as if too hot to handle. When I came downwind of the activity, I realised why the hyenas were so reluctant to get too involved. The polecat was letting off the most awful, strongly pungent smell. Finally it escaped down a hole. I wonder why the one Bop attacked did not also use this chemical defence. Perhaps Bop caught it by surprise before it had the chance to do so.

Having watched brown hyenas eat ostrich eggs with such ease, I was surprised to find that spotted hyenas have so much difficulty with this pursuit. The first time I saw this was with Old Flat Ear. She came across a single egg lying in the sand, but try as she might she simply could not get her mouth round it to open it. I thought that her worn teeth had perhaps hindered her.

Sometime later I followed six hyenas to an ostrich nest containing 15 eggs. Unlike the occasion when the brown hyena Cicely found a deserted nest, this time the cock was sitting on the eggs and tried to distract the hyenas by running off while performing the 'broken wing' display. This did not fool the hyenas and they went straight for the eggs. However, they also struggled to bite them open. Being spotted hyenas, they had an ingenious solution to the problem. They repeatedly tried to kick the eggs backward with their forefeet. Although most kicks missed the eggs, sometimes they connected and the egg shot backwards. Even more rarely, one of the eggs collided with another and a distinct crack was heard. At this the hyena would bite at the cracked egg and easily open it. In this way they cleaned out the nest. I am at a loss to explain why brown hyenas open ostrich eggs so easily, yet spotted hyenas and even lions are not nearly so capable. It can't have anything to do with the size of their gapes, as the brown hyena is the smallest of the three.

Spotted hyenas are able to take in copious amounts of food, and are presented with more opportunities to do so than are brown hyenas. When they are really full their stomachs bulge, fit to burst. Often people informed me that they had seen an extremely pregnant hyena, when in reality it was an extremely full one. A pregnant spotted hyena at full term only carries an extra three kilograms or so in her stomach, while a satiated hyena may carry nine times as much. In addition they are able to repeat the performance at remarkably short intervals if circumstances permit. For example, in a five-day period Olivia ate the following: one springbok lamb (5 kilograms), a springhare (2 kilograms), a quarter of an adult springbok (9 kilograms), a tenth of another springbok (4 kilograms) and a sixth of a gemsbok (20 kilograms) – a total of about 40 kilograms of meat, or 8 kilograms a day, not to mention the few bones she chewed on in between the meatier items. At other times of course they are not so fortunate. I once followed four young males from the Kousaunt Clan for three nights, during which time they covered 118 kilometres and only managed to find a few old bones to chew on.

As with the brown hyenas I tried to work out an average daily consumption rate for spotted hyenas. I arrived at a figure of 6.2 kilograms a day, made up of 4.5 kilograms from kills and 1.7 kilograms from scavenging. This is a very high figure for spotted hyenas. Hans Kruuk found that his spotted hyenas in the Ngorongoro Crater consumed 5.4 kilograms a day during the bonanza provided by wildebeest calves in the calving season. At other times of the year the figure dropped down to only 2.0 kilograms a day. Kruger hyenas, according to Joh Henschel, make do with 3.8 kilograms a day.

This result is puzzling. If Kalahari spotted hyenas do have so much food available, why are there not more of them? Perhaps, the large distances they are forced to move on most nights to find their food – usually more than 30 and sometimes as much as 70 kilometres – require a larger amount of energy than their Ngorongoro and Kruger counterparts, which rarely travel more than 20 kilometres a night. In other words, perhaps the energy budget of Kalahari spotted hyenas is not as well balanced as it seems.

Like brown hyenas, spotted hyenas have a wonderful sense of smell. They are also, I believe, several notches up the intelligence scale than brown hyenas and indeed the other large African carnivores: lions, leopards, cheetahs and wild dogs. An illustration of both their fine sense of smell and their intelligence is provided in the following incident.

I was at the Kousaunt den with Tu-Tu and two of the young males from the clan. It was well after midnight and none of the hyenas had shown any inclination to move. It looked like becoming a long, boring night with the BBC World News repeated six more times. However, by then I had learnt well the lesson that when watching animals, patience is not a virtue, but is crucial. On this particular night my patience was rewarded when Old Flat Ear came to the den with an extended belly and a face covered in blood. The three hyenas at the den moved over to the matriarch, showing her the deference her rank demanded. Only Tu-Tu was confident enough to come right up to her and sniff her bloody head. Tu-Tu then moved away from the den, her nose to the ground, seeming to follow the trail along which Old Flat Ear had come.

After going 500 metres she turned round and ran back to Old Flat Ear, once again sniffing her around the head. She then lay down but remained restless, often lifting her head and looking over to Old Flat Ear, who was now suckling her small cub. After about 10 minutes Tu-Tu stood up and again sniffed the old female on the head. Then she started moving away from the den, again apparently following Old Flat Ear's trail. This time the two young males that had been watching these comings and goings followed Tu-Tu. The three hyenas crossed over a large plain, moving quickly with their noses to the ground.

Every now and then they would circle, obviously having lost the scent, before picking it up once again and continuing along the scent trail left by Old Flat Ear. Eventually, eight kilometres from the den, they arrived at a freshly killed wildebeest carcass on which Hans was feeding.

From what I have described both in this chapter and in chapter 4, it is obvious that there are large differences in the diets and foraging habits between brown hyenas and spotted hyenas. Put in a nutshell, the brown hyena is a solitary scavenger, feeding off small food items that usually provide only enough food for one hyena at a time, whereas the spotted hyena is a hunter-scavenger that feeds on large food items which can provide a meal for several hyenas at a time. This largely explains how these two closely related animals can coexist: they do not compete very much for food. It is also the basis for the large differences in behaviour between these two species.

One striking difference between the brown hyena and spotted hyena is that the former walks around on its own, whereas spotted hyenas frequently forage in groups. In the Kalahari spotted hyenas are mostly encountered in groups of three to five. We have seen that brown hyenas do not need to co-operate to obtain their food and, more importantly, most of their food items provide a meal for only one hyena. The obvious reason why spotted hyenas hunt in a group is that in order to overcome large prey like gemsbok there is a need for the hunters to co-operate. This, however, is only part of the answer. As we have seen, a solitary spotted hyena, for example, pulls down gemsbok calves as easily as three or four do hunting together. The main point is that when they do pull down a gemsbok calf, unless it is very young, there is enough food to feed several hyenas. The animals that share a carcass are not arbitrarily chosen but are selected by rather stringent rules of hyena society. However, before I discuss this I would like to show where the spotted hyena fits into the community of antelope hunters in the Kalahari.

CHAPTER 9

Spotted Hyenas
versus the Cats

EACH SUNSET that Hermanus and I set out to look for hyenas there was an air of expectation – we could never predict what would happen before the sun rose again. Often, of course, very little would happen and we would spend several boring hours watching hyenas sleep and listening to the BBC. Other nights were packed full of excitement and interest that kept us going until after sunrise.

One such night occurred in May 1980, although it started quietly enough. We came to the Kousaunt den to find Old Flat Ear lying there alone. She did not appear to be interested in anything but sleep that night. Not knowing where any other members of the clan were, we had no alternative but to take our chances with the old female. At 9 pm she stirred, but after a good stretch she only moved off a few hundred metres, then flopped down again. Although five minutes later she was up, she really did not appear to have much enthusiasm and was soon down again. In fits and starts she gradually moved into and then across the Nossob river-bed and into the dunes on the Botswana side. By 11.30 pm she had moved only eight kilometres when her behaviour changed drastically.

Suddenly she began loping with her nose in the air, obviously having smelt something interesting. After two kilometres she stopped and looked off intently into the night. A careful scan with the spotlight revealed a lioness some 100 metres away, lying next to an only slightly eaten gemsbok carcass.

Old Flat Ear stood staring at the lioness for several minutes. She moved closer and stopped again, her eyes fixed on the lioness and her kill. Then she started to whoop – the characteristic call of the spotted hyena. Old Flat Ear had a characteristically deep guttural whoop, which she usually repeated six or seven times in a bout. This night she whooped 17 times consecutively. Almost immediately a hyena replied and within two minutes Olivia, Hans and Old Flat Ear's one-year-old son Six Darts (so called because that's how many it had taken to immobilise him for marking – I missed five times!) came running up to her. Did she know they were in the vicinity or was it her good fortune? I think the former.

After noisily greeting each other, the hyenas approached the lioness, with the three adults advancing shoulder to shoulder and the younger one bringing up the rear. Until the reinforcements arrived the lioness had paid very little attention to Old Flat Ear. Now she dragged the carcass deeper under the tree and started feeding. It was as if she knew that her time at the banquet table was nearly up. The hyenas, emitting the most intimidating series of lows, hoot-laughs and whoops, kept on approaching, their ears cocked, their black tails bristling and curled over their backs. The lioness sat up and faced them, trying to intimidate the hyenas by growling fiercely, but her posture – ears flattened and teeth bared – showed a far more defensive type of aggression than the confident attacking postures of the hyenas. When the hyenas were a metre from her, the lioness lunged at them, giving vent to a loud roar-growl, in a last desperate attempt to retain her hard-won prize. After only a flicker of hesitation the hyenas closed in. At last the lioness's nerve broke. She spun around and, crashing through the tree's branches, beat a hasty retreat, leaving her carcass to the hyenas.

Hermanus and I were impressed. Who, I thought, would believe that these so-called cowardly scavengers could summarily dismiss the regal lion. I had heard of such behaviour from Hans Kruuk, but this was the first time I had witnessed it.

In the southern Kalahari lions and spotted hyenas do not meet up very often – I witnessed only 33 interactions between these two arch-rivals during the more than 1 500 hours I spent watching spotted hyenas. These rare encounters more often than not escalated into a dramatic encounter and provided some of the most memorable scenes of my entire study. Although a dispute over food was often the reason for an altercation, it was not always so. When food was not the cause, it was the hyenas that usually seemed to provoke the situation. Sometimes they went far out of their way to engage lions.

One example of this was when six hyenas we were following came to Kousaunt Windmill. Instead of stopping to drink and take time out for a bit of a rest, as was their custom, they loped straight past the windmill without so much as taking a lick of water. Ahead in the moonlight, I could make out four silhouetted gemsbok, but Hermanus and I – and I imagine the gemsbok too – were rather surprised when the hyenas ran right past them without a glance in their direction. Obviously there was something else that had caught their attention.

Exactly 2.7 kilometres from Kousaunt they found it. Suddenly the hyenas darted to the right and gave chase. I caught a glimpse of a large animal taking off. Two hundred metres further on, the hyenas stopped at a dead tree trunk and halfway up, clinging on for all she was worth, was a large lioness. Within seconds the hyenas dashed off again after a second lioness. She managed to maintain her composure rather better than her companion had and stood her ground, growling. With the hyenas facing her in a tight bunch, the cat lunged at them and they jumped back, before regrouping. By now the first lioness had climbed down and joined the second one. The cats rubbed heads in greeting, surely pleased to be together again. They attempted to move off, but this excited the hyenas, who advanced on the lions whooping and cat-calling. For the next hour and a half the hyenas kept the lions there. If the lions stayed still, the hyenas were content to let them lie. However, whenever a lioness sat up or tried to move, it invited harassment. Eventually the hyenas tired of their siege of the lions and moved away.

Sometimes, of course the shoe was on the other foot. This was particularly so when adult male lions were involved. Then the hyenas treated the cats with far more respect. Once, Olivia and her two seven-month-old cubs were feeding on the remains of a springbok when a large male lion bounded up to them and caught one of the cubs as it ran away. With one shake the cub was dead.

I attempted to score the results of spotted hyena–lion clashes in order to measure the effect that these had on each species. Firstly, I asked which species deprived the other of most food and, the opposite question, which species provided the other with most food. Surprisingly the scores were pretty equal. Of the 13 clashes I observed over food, spotted hyenas lost their kill to lions four times each and vice versa. Twice lions and thrice hyenas managed to maintain possession of the carcass when challenged by the other species. Of course lions also provide spotted hyenas with food in the same way as they do for brown hyenas, when they leave skin and skeleton from kills. In these instances spotted hyenas come into competition with brown hyenas. This is another kettle of fish which I will describe in another chapter.

Away from food the situation is a little different. Lions initiated a skirmish seven times and hyenas did so nine times. Four times I could not decide who started the conflict. Lions were judged to be winners ten times, hyenas won only five contests – in three of them the lions ended up climbing trees to escape from the hyenas – and five ended in a draw. What was perhaps most striking was that when hyenas initiated the skirmish it lasted about an hour, whereas when lions did so it lasted only about five minutes.

Why should this animosity between lions and spotted hyenas exist? To answer this question we have to look at the feeding habits of the two species. For both, gemsbok and wildebeest make up nearly 70 per cent of their kills. If they are competing for a scarce resource, and in the Kalahari all resources are at times scarce, it is not in the one species's interest to have the other species in the vicinity. This is particularly so for the smaller competitor, the spotted hyena. If lions come into an area occupied by spotted hyenas and the spotted hyenas make life as difficult as possible for the lions, this may encourage the lions to vacate the area and move on.

Lion mobbing is obviously a dangerous occupation for spotted hyenas to indulge in. The stakes are high and a false move or indiscretion can lead to death. In the Ngorongoro Crater, Hans Kruuk found that hyenas were far less prepared to take on lions than their Kalahari counterparts were. There, resources in the form of wildebeest and zebra are far more abundant than they are in the Kalahari. If hyenas should lose a carcass to lions in this rich area they can fairly easily go out and kill another one. In contrast, Kalahari hyenas might have to travel over 50 kilometres before locating the next gemsbok herd.

Lion mobbing brings out the co-operative spirit in spotted hyenas more than practically any other activity. Whereas the average foraging group size for spotted hyenas in the Kalahari is three, the average lion-mobbing group size is almost seven. Lion mobbing is always accompanied by a rich cacophony of sounds. Apart from getting the hyenas 'psyched' up and intimidating the lions, these noises may also serve as a rallying call. Certainly, any hyenas in the vicinity come running in to assist their colleagues when lions are under the spotlight.

Apart from these high-profile contests, there are other more subtle ways in which lions and spotted hyenas avoid competition. The most important is the manner in which they select their prey. Although gemsbok and wildebeest are the top prey for both, hyenas catch mainly calves, whereas lions concentrate on adults. This suggests that each species has adapted its hunting behaviour to different segments of the prey population, helping make it possible for them to coexist in the Kalahari. Of course they have not deliberately done this: it has been through the process of evolution and natural selection.

The other two big cats that inhabit the Kalahari, the leopard and the cheetah, also come into contact with spotted hyenas from time to time. As these two species are mainly hunters of springbok and smaller prey, they do not compete for food with spotted hyenas to the extent that lions do. Consequently, spotted hyenas do not show the interest in these two smaller cats that they do in lions. Furthermore, the leopards' habit of taking their kills into trees and the cheetahs' mainly daytime hunting activities further diminish the occasions for hyenas and these cats to clash. Nevertheless, there are times when they meet and sometimes there is a take-over of kills.

Hermanus once skilfully reconstructed an interaction between hyenas and a leopard from spoor. Evidently the leopard had killed a gemsbok calf and dragged the carcass some 150 metres to a shepherd's tree. This was in an area where trees of any description were few and far between; moreover, the shepherd's tree was actually unsuitable for dragging a kill into, being more of a bush. Consequently the leopard could only drag the kill under the low hanging branches. While the leopard was eating, the three hyenas appeared and chased it away, devouring the remains of the carcass, except for the head and horns. When the hyenas departed, the leopard returned and carried off the remains to another small tree 300 metres away and placed it in its top. Too little too late!

In my experience leopards always give way to spotted hyenas, even in a one-to-one encounter. Koos Bothma, who together with Elias le Riche and their trackers studied leopards in the Kalahari for several years, recorded two instances of large male leopards successfully defending their kills from a spotted hyena.

Cheetahs are built for speed more than for strength. By hunting mostly during the day they avoid competition from stronger competitors. This was vividly brought home to me one moonlit night when three cheetahs killed a springbok and were immediately chased off the carcass by five spotted hyenas. About two hours later the cheetahs again killed a springbok and just as quickly the hyenas robbed them of their prize again.

Early one morning in the St John's Dam region, Hermanus and I were looking for spotted hyena spoor when we came across a female cheetah and her four almost independent offspring. While we were watching them, they flushed out and easily caught a steenbok. Within minutes four hyenas came running up and claimed the remains of the cheetahs' kill. As expected, the cats surrendered the food without protest and moved off 100 metres to the top of a dune where they lay down. Once the hyenas had finished the remains they moved slowly towards where the cheetahs were lying, with their noses to the ground. Although they could obviously smell the cheetahs, it appeared as if the hyenas did not see them. As they approached, the cheetahs lay flat watching the hyenas, as if hoping that they would not be discovered. I was surprised how close the hyenas approached before finally seeing the cheetahs. When they did the cats bolted. Until they took refuge in two large camelthorn trees with remarkable agility, it looked as if it would be a rather desperate situation for the cheetahs. The hyenas were stymied by the manoeuvre and after about 10 minutes moved away.

Both the cheetah-killed springbok that provided brown hyenas with so much food along the Auob river-bed and those killed in the Nossob were the source of a useful set of data. Fortunately, except in the case of lambs, the skull is the last part of a skeleton to disappear. Even hyenas are unlikely to break this up, at least not the parts I was interested in, the teeth and horns. By examining the bottom jaws of these kills and noting the state of wear of the teeth, I was able to classify each jaw into one of five age classes. By comparing the age distribution of this sample with a sample of jaws from springbok which were shot for rations, and which represented the age structure of the living springbok population, I found that the cheetah-killed sample consisted of many more old animals than the shot sample. This shows that cheetahs are more likely to kill old springbok than those in their prime. By examining the horns and noting the sex of the kill, I also found that many more males were killed by cheetahs than females. Cheetahs, therefore, tend to kill the most expendable members of the springbok population – old males – which means that they have less of an impact on this population than they would if they killed females in their prime. A similar analysis of wildebeest killed by lions and those that died from causes other than predation showed that they too tend to kill the most expendable members of the population, in this case those that are doomed to die from disease or starvation.

The relationship between predators and their prey is a very complex one and it is certainly not a simple matter of predators keeping prey numbers down. In the Kalahari ecosystem it appears that predators have little impact on their prey populations in the long term when drought seems to be the main cause of death for most species. This is a result not only of the way in which predators select their prey, but also due of the nomadic nature of the prey compared with the more sedentary behaviour of the predators. At times there are few prey animals available to the predators on the South African side of the park: they move out of the area into the Botswana section of the ecosystem where rainfall is higher. The predators, particularly the cats, are unable or unwilling to follow these nomads, especially when they have young. This sets a limit on the number of predators that can survive, as food becomes scarce and starvation becomes a reality.

An exception may be the gemsbok, which is the least nomadic of the larger prey species and which suffers from a combined onslaught from spotted hyenas and lions, as well as losing young to leopard and cheetah. It is interesting that the density of gemsbok in an area bordering the park, with comparatively few large carnivores, was twice as high as it was in an area of similar habitat within the park with the full complement of large carnivores. I do not know if predation was the only, or even the main, reason for this – it might have been due to the fact that the gemsbok in the area outside the park were compressed into a smaller area because of human pressure – but it does suggest that predators played a role. Another exception is the steenbok. These cryptic little antelope are numerically the most abundant antelope species on the South African side and are an extremely important prey species for cheetahs.

Although predation may not be the main factor limiting prey numbers in the Kalahari, predators certainly influence the structure of prey populations. For example, the distorted sex ratio in springbok, the females of which outnumber males by nearly two to one, is mainly a result of cheetah and other predators killing more males than females. Predators, too, may have had an influence on the evolution of some of the attributes of their prey. The vigilance of the hartebeest, the agility of the springbok, the stamina of the wildebeest and the rapier-like horns of the gemsbok – characteristics which we enjoy and admire – have all evolved to help animals escape from predators.

The gemsbok example is especially relevant here. In general, antelope horns have not evolved as a means of defence against predators. If this was so, then females would be as well endowed as males, yet in many species horns are less well developed or even absent in females. Furthermore, the twisted shapes of many species' horns are not really useful as weapons of defence. Rather, horns are an example of sexual selection and are mainly used by males as a display to attract females and intimidate rivals, like the lion's mane. In the case of gemsbok both sexes have equally long horns and their rapier shape makes them very useful as stabbing instruments. I also discovered that horn growth in young gemsbok is far more rapid than in other antelope. As gemsbok calves are especially vulnerable to spotted hyena predation, it makes sense for them to grow their horns as quickly as possible to increase their chances of survival.

CHAPTER 10

The Battles of the Sexes

FOR A LONG TIME spotted hyenas (but not the other species of hyena) were considered to be hermaphrodites, even though the great Aristotle had refuted this thousands of years ago. As recently as the 1950s a keeper of spotted hyenas maintained that an animal that for years had been a male changed sex and produced young. It is not hard to understand how this myth arose, for it is extremely difficult to tell the sex of a spotted hyena – they all look like males. The female mimics the male's reproductive organs. She has a false scrotum, which consists of the fused vaginal labia and a greatly enlarged clitoris that looks for all the world like a penis and possesses the same erectile properties. Even small cubs possess these prominent sexual organs. In spite of these confusing anatomical features it is not so difficult to tell the sex of a hyena provided the animal obliges and erects the organ. The male's penis normally points backwards and terminates in a pointed tip. The female's clitoris is likely to point forward and is blunt at the tip.

Even more remarkable than the shape of the female's reproductive organs is the fact that this structure houses the birth canal. Female spotted hyenas are the only mammals that lack an external vaginal opening. Mating and birth take place through a urogenital canal at the tip of the clitoris. This poses problems for the female when giving birth as the clitoris tears to allow the baby to emerge, leaving a wound that takes several weeks to heal. First births may even cause the death of the baby. A female that has given birth can thereafter be identified by a vertical band of pink scar tissue along the clitoris. It is also a challenge for the male when mating, as the opening in the clitoris points forward and is not erect during copulation.

For many years a team of scientists lead by Laurence Frank and Steve Glickman at the University of California at Berkeley have researched this weird phenomenon. This has proved to be a daunting challenge as it is a complicated process involving the combined effects of a number of hormones. At this stage it is not fully understood, but female spotted hyenas do possess inordinately high levels of the male hormone testosterone, which is necessary to maintain the aggression needed for their dominance.

The strange sexual organs also feature in the bizarre meeting ceremony of the spotted hyena. When two hyenas from the same clan meet they stand head to tail, lift the hind leg nearest to one another and, often with a deep lowing sound, sniff and lick the erected sexual organs of the other. In effect the two meeting animals are displaying their most vulnerable parts to the other's most lethal ones: its teeth and jaws. This entails a large amount of trust and the ceremony will only be indulged in by two animals who share this mutual trust. The erect sexual organ is a sign of submission, so it also helps to lessen the amount of aggression in the clan. Cubs are always keen to engage in the meeting ceremony and are able to erect their penis or clitoris as young as four weeks of age.

During the first year of my observations of the Kousaunt Clan I began to detect differences in the behaviour of the members. The most strikingly different hyena in the clan was the adult male Hans. He was very much the odd hyena out when it came to the social activities of the clan. He did not feel comfortable enough with the other members of the clan, particularly the four adult females, Old Flat Ear, Olivia, Ella and Tu-Tu, to present to them in the meeting ceremony. In fact he avoided direct contact with them and would jump away if a female approached him. He was also often missing from a foraging group and spent far more time on his own than any other members of the clan. When he did join some of the others at a carcass, he was usually chased away and not allowed to eat until all the others had had their fill. This is not the way I imagined that the only adult male in the clan would have been treated.

During the second year his status improved somewhat and, although he would not perform a meeting ceremony with an adult female, he sometimes did so with some of the younger members of the clan. He also began to appear far more often in foraging groups with the others.

In other ways Hans's behaviour made far more of an impact. Being in the presence of the other members of the clan, particularly the adult females, was obviously important to him. He whooped more often than any other member of the clan and indulged more frequently in the strange hyena scent-marking behaviour known as pasting. He also often took the lead in foraging and hunting, although he was still very much a second-class member of the clan when it came to the sharing of the spoils.

His behaviour in fact was typical of that of an immigrant male rather than of a natal male. As I look back at the fight between Hans and the unknown male at Groot Brak the night I caught most of the Kousaunt Clan, it seems likely that this was a fight between two prospective immigrant males. In the year after the fight I twice saw the loser, once near Nossob Camp, 40 kilometres from Groot Brak, and then near Dankbaar, 70 kilometres from Nossob. Both times he was on his own, although the second time he was close to members of the Leijersdraai Clan. Hans stayed in the area and, as I have described, slowly became accepted by the other members of the Kousaunt Clan.

For two years Hans enjoyed his position as the only adult male member of the clan. The young males that had been present in the clan in January 1979 gradually disappeared. In fact during my entire study all males disappeared from their natal clans at about three years of age. I was unable to follow the fate of most of them, but at least three were nomadic for several years and another, Olivia's son Silver, successfully joined the St John's Clan down near Twee Rivieren. In May 1981 a second adult male called Necklace, because of a series of spots around his neck, appeared on the scene. For the next two years he was peripheral to the Kousaunt Clan – often in the background, but never managing to integrate into the clan like Hans had. Hans, in fact, was the chief force, although well supported by the adult females, in keeping Necklace away from the other members of the clan.

I had to admire the persistence of Necklace: over a year after his first appearance he was still very much an outsider. For example, once we followed nine hyenas from the Kousaunt Clan on an extended foraging trip of over 60 kilometres in two nights. For the entire journey they were shadowed by Necklace. From time to time he would make an appearance, only to be chased away by Hans or some of the other clan members.

Why should male spotted hyenas spend so much time and energy trying to be accepted into a clan? Immigrant males are the ones responsible for mating with the females. For his persistence Hans was rewarded with a reign of over six years as the alpha male of the Kousaunt Clan, during which time I believe he sired 23 offspring. (Unfortunately in those days it was not possible to collect DNA in order to prove paternity.) As we shall see later, Necklace was also eventually rewarded for his persistence.

In a male 'winner takes all', or polygynous, mating system as is found among spotted hyenas, it is usual for males to display a large amount of aggression towards each other and to be adorned with extravagant plumages for displaying to females. Amongst carnivores lions are the classic example of this, but spotted hyenas do not fit this model. The males have no adornments and are even smaller than their mates, and they are not particularly aggressive towards one another. The fight between Hans and the stranger at Groot Brak, and the many skirmishes I watched between Hans and Necklace, were tame in comparison with the fight-to-maim or -kill battles that occur between competing male lion coalitions.

Rather than the macho characteristics of male lions, persistence seems to be an important quality for male spotted hyenas to possess. They need to reinforce their presence in the territory by whooping and scent-marking, following the females and, perhaps most important of all, taking the lead in hunting. It seems as if the females have a say in which males are allowed to join the clan, rather than the males settling this among themselves. Spotted hyena society is strongly female-dominated.

The role of males in spotted hyena society is somewhat different from that in brown hyena society. There are two major reasons for this: the dominance of spotted hyena females and the fact that males do not act as helpers by bringing food back to the den. Spotted hyena cubs drink mother's milk and practically nothing else for the first nine months of their lives. Because males are unable to enhance their reproductive fitness through kin selection, by helping to feed the cubs, there is no point in a male spotted hyena staying with his natal clan once he has grown up. He must take his chance in the outside world. Being nomadic does not hold the same reproductive opportunities for male spotteds as it does for male browns, so every spotted hyena male's ambition must be to become accepted by a group of females. In the small Kalahari clans this practically guarantees good mating opportunities.

All this raises the question: why don't spotted hyenas feed their cubs by carrying food to the den? I believe that the answer lies in their feeding habits. Because spotted hyenas mainly feed on food items that can provide a meal for several hyenas simultaneously, such as gemsbok calves, they often forage and feed in groups. At carcasses there is competition for food, which is expressed in the speed at which the hyenas eat. This is in striking contrast to the leisurely manner in which brown hyenas usually feed. Should a spotted hyena wish to carry a piece of food off to the den, it would have to do so almost as soon as feeding began, and by the time it returned to the carcass all the choice bits would have been eaten. Furthermore, in contrast to brown hyena dens, where there are usually only one or two litters of cubs, there are likely to be several litters with cubs of different ages at a spotted hyena den, and the larger cubs would take the bigger share of any food brought back to the den.

The males are overshadowed in spotted hyena society by the larger, macho, testosterone-charged females. When the territory is under attack from neighbouring clans, or when lions are to be challenged, it is the females that stand shoulder to shoulder and lead the troops into battle.

In the early years of my association with the Kousaunt Clan, I noticed little aggression or competition between the four adult females of the clan and their offspring. Old Flat Ear was obviously the matriarch of the clan, although I did not know how the other females were related to her and to each other. While I was away in Scotland between February and May 1981 writing up the brown hyena work for my PhD, Old Flat Ear died, and when we returned, several changes had taken place in the clan. Olivia had two small cubs, her oldest daughter Goldie had produced a single cub, and her other daughter Guy (so called because for a long time I thought she was a male) was now a sub-adult. Ella also had two tiny cubs, and her male and female cubs from the previous litter were nearly two years old, although both disappeared a few months later. Finally, Tu-Tu had a male and female cub of about one year, which I called Joh (after Joh Henschel, who was about to start a study of spotted hyenas in Kruger National Park) and Soap (for want of a better name) respectively. It was also at this time that Necklace was first seen hanging around in the background.

Over the next few months changes in the behaviour of the females towards each other began to occur – there was more conflict between them. In particular Olivia and Goldie, with Guy often in attendance, were intolerant towards other members of the clan, except their own offspring. A typical interaction occurred when Olivia and Soap (Tu-Tu's daughter) arrived at a gemsbok carcass on which Guy was already feeding. Olivia immediately started to feed, but Soap showed no inclination to do so and stood some distance away looking on. After a few minutes she approached the carcass only to be attacked immediately by Olivia and driven off. Next, Goldie arrived with her 14-month-old cub and one of Ella's cubs of about the same age. Goldie and her son were allowed to feed, but Ella's son joined Soap on the periphery. When Soap's sibling, Joh, arrived, he too was excluded from the feast. The only animals that were allowed to feed from this carcass were Olivia and her descendants.

As time went on and the number of hyenas in the clan increased, this trend became stronger. When Guy also produced her own offspring and as Goldie's cubs grew up, Olivia and her relatives began to form a powerful sub-group or coalition. Tu-Tu produced a new litter of cubs, though her grown-up daughter Soap failed to do so. Ella's son and daughter disappeared from the clan. A clear dominance hierarchy began to develop with Olivia as top hyena, closely followed by her daughter Goldie. Ella was dominant towards Tu-Tu, but was dominated by Olivia. Guy was dominated by her mother Olivia and older sister Goldie but avoided conflict with Ella and Tu-Tu.

The clan had now divided into three coalitions, each headed by one of the three surviving original adult females, Olivia, Ella and Tu-Tu. Each coalition comprised the head female and her offspring. Whereas prior to 1983 a foraging group from the Kousaunt Clan might comprise any combination of members of the clan, after this date most foraging groups were made up of members from one coalition only.

In early 1983 the number of adult females in the clan reached a peak of eight. The Olivia coalition was the largest with nine members, the Tu-Tu coalition had five and the Ella coalition four. The atmosphere in the clan was tense and, without being able to put my finger on it, I sensed that something dramatic was happening. I decided to spend as much time as possible with the hyenas. Whenever I could get away from other duties, irrespective of the moon phase, I would travel up to Kousaunt to check on their activities.

On one such night in late July my mother, who was visiting from Johannesburg, and I witnessed a very dramatic and important event. As usual, we started our observations by visiting the den at sunset. Olivia and her entire coalition and Ella's six-month-old cub were present. Soon after we arrived the entire entourage, including Ella and Guy's singleton cubs and Goldie's two cubs, started moving off to Kousaunt Windmill about two kilometres away, leaving no hyenas at the den. Both Olivia and Guy were limping. Guy in particular was dragging her back legs – almost certainly injuries sustained in a recent fight. When they arrived at the waterhole they all had a drink and lay down. For the next 10 minutes all was quiet. This was the lull before the storm.

Suddenly, like fighter planes making a surprise attack, six hyenas charged in from the dark. The Olivia hyenas scattered, the cubs retreating, while the older ones quickly regrouped to face the onslaught. After this it was difficult to keep an account of exactly what was happening. The two groups went hammer and tongs at each other in a bloody and dusty battle, accompanied by an incredible cacophony of whoops, roars, growls, yells, hoots, squeals and giggles. They bit each other around the face, chest and stomach, the attacked animals going down onto their stomachs to protect their vulnerable underparts. At one stage one of the hyenas took refuge under my vehicle, only to be pulled out again by the others. I was unable to identify the attackers but noticed one of them standing to one side, watching the action. I quickly recognised it as the male Necklace. Also standing about 30 metres away from the action and whooping were the four cubs from the Olivia and Ella coalitions.

Nine minutes after the original attack, the unknown hyenas broke away and ran off up the Nossob river-bed with the Olivia group in hot pursuit. I raced off after them in the vehicle as it was imperative to know who was responsible for this very unusual attack. After about one kilometre Olivia and her compatriots gave up the chase and after another 2.4 kilometres the six attackers eventually stopped and lay down. Immediately I picked up the characteristic ear markings of Tu-Tu. Four of the others were her offspring from her two previous litters. All of them, but particularly Tu-Tu, wore red battle masks of blood. The sixth hyena was Necklace, who was unscathed.

Was the Battle of Kousaunt the climax of a gradual separation between these two coalitions? I stayed the rest of the night with Tu-Tu and company. When they moved on I was immediately impressed by the behaviour of Necklace. After two years as a peripheral nobody, constantly being chased away by all and sundry, he now took up a position in the centre of the group. He walked close to Tu-Tu and, when they visited six latrines that night, he, just as Hans always did, made the major scent-marking contribution – pasting his anal gland secretions onto grass stalks and then vigorously scraping the ground with alternate forefeet. His behaviour had changed practically overnight from that of a peripheral immigrant male to one of a central immigrant male.

The next evening I was surprised to find Tu-Tu and her family (the defeated team) with Necklace at the den. None of the Olivia coalition members were present, but lying about 50 metres away were Ella and her cub, which had been present at the fracas the night before but had taken no part in it. The two groups ignored each other and soon after sunset the Tu-Tu group moved off. I followed them for 11 kilometres as they moved east into Botswana. There they scavenged a hartebeest carcass. Once again the fundamental principle of spotted hyena society was expressed as Tu-Tu and her descendants fed from the carcass while Necklace hovered in the background.

I had to get back to Nossob the next day. When I came out a few days later I found no trace of the Tu-Tu coalition. In fact I never saw these animals in the Kousaunt territory again. It was not until December 1984, over a year later and after we had left the Kalahari for Kruger, that Mike Knight found Tu-Tu and her associates, including Necklace, at a den about 40 kilometres to the south-east of the Kousaunt den. They had obviously set up their own clan territory in this region.

In 1986 Mike Knight recorded that Ella and a daughter disappeared from the Kousaunt Clan, although he was unable to find out what happened to them. The Kousaunt Clan, therefore, eventually contained only animals that were descended from Olivia, except of course for Hans. Olivia and Hans died of rabies in 1987, leaving Goldie as the matriarch and giving way to a new central immigrant male.

Female spotted hyenas in the Kalahari appear to have three choices when it comes to spreading their genes. Number one choice is to remain with the natal clan and raise their own cubs at the communal den. However, not all cubs can do this as there is a limit to the number of hyenas, and particularly breeding females, that a clan territory can feed. The daughters of the dominant female are born into a privileged position and are most likely to inherit a place in the clan. Unlike the males, female spotted hyenas are able to practise kin selection by forming breeding coalitions with their offspring and by allowing close relatives to feed on carcasses and excluding more distantly related members.

Less dominant females are less likely to be able to breed in their natal clan. They have to be more ingenious than those that dominate them. One strategy they can employ is to build up a coalition of non-breeding offspring, then emigrate to start a new clan. The success of this strategy will depend on the availability of a vacuum territory in which to start the clan.

Finally, on two occasions I documented the very unusual occurrence of a single female leaving one clan and joining another. Normally clan females are strongly intolerant of strange females. Unfortunately I was unable to study these clan transfers in any detail to establish what unusual circumstances might have been present in the clan to make this possible.

Spotted hyena females invest much energy in bringing up their cubs – more than any other carnivore. Milk production is very costly in terms of energy and the females have a very long lactation period, when their cubs' only source of nutrition is their mother's milk. Furthermore, they produce milk with the highest protein content recorded in any terrestrial carnivore and a fat content only exceeded by polar bears and sea otters. This must surely explain why spotted hyenas never raise more than two cubs per litter. It may also explain why spotted hyena females tend to be larger than males and dominant to them. The larger you are the more energy, in this case in the form of milk, you can produce. This is known as the big mother hypothesis – big mothers make good mothers. Moreover, the larger and more aggressive you are, the better your chances of being able to obtain a larger slice of the cake in order to produce the required energy.

The importance of dominance in spotted hyena society was nicely illustrated in the Kousaunt Clan. Dominant females produced more cubs than the others: Olivia produced eight cubs during the study, whereas Ella produced seven and Tu-Tu only six. What is even more important, Olivia had nine descendants over two years of age at the end of 1983, whereas Ella and Tu-Tu only had five each.

CHAPTER 11

Other Clans
and Other Lands

BESIDES MY DETAILED STUDY of the Kousaunt Clan, I needed to get to know some of the other spotted hyena clans in the Kalahari. Not only did I want to get an idea of the number of spotted hyenas on the South African side of the park, but I also wanted to learn whether spotted hyena numbers, like those of the brown hyena, are influenced by food availability. The important things to look at were the size of some spotted hyena territories and the number of members in each clan At the same time the tricky twin questions of the amount of food available and the manner in which it was distributed in each territory needed answering.

The first of the new clans I looked to was the Kaspersdraai Clan. Way back in 1974 I had caught and ear-notched some of these hyenas. During the late 1970s they disappeared and it was only in 1981 that we began to see hyenas in this area again. There was an adult female, whom I called Marlin, an adult male called Cyclops and Marlin's cub of one year, Agfa. Marlin was named after Margie, whose second name is Elin, Cyclops was blind in one eye and Agfa was named to celebrate my luck in winning the Agfa Wildlife Photographic competition that year. Cyclops limped badly on his left front foot, almost certainly the result of a gin trap from which he had escaped.

Towards the end of 1982 we also began to find signs of spotted hyenas around Nossob Camp for the first time in many years. We soon established that the animals responsible for these signs were a clan of eight hyenas that denned near Seven Pans, approximately 25 kilometres west of Nossob Camp. The matriarch of this clan was an old female with a large growth on her back whom we called Lumpy Back.

A third clan I started looking at in 1983 was one down in the southern Nossob near what used to be called St John's Dam but today is called Kijkij. I decided to persist with the old name as the school I went to in Johannesburg was St John's. The St John's Clan had first caught my attention in 1976 when three of the females I had marked two years earlier at the Kaspersdraai den some 75 kilometres away appeared in this area. One of them was still with the clan when I started to make more formal observations in January 1983. At this time the St John's Clan comprised 13 animals. Also of interest was that the central immigrant male was Olivia's son Silver. He had settled down over 120 kilometres from his birthplace.

As my main objective with these clans was to measure the size of their territory, rather than to make detailed observations of behaviour, we spent little time watching the hyenas. Instead, we reverted to a more diurnal existence and took to tracking their spoor in order to learn where they had moved and what they had eaten the night before. With Hermanus perched on his tracking seat, on a good day we could follow the tracks across the dunes at a speed of 20 kilometres an hour. Of course the hard river-bed was still a problem and often held us up, or caused us to abandon a particular set of tracks. Nevertheless, over the next 14 months, from my vehicle or on foot, we followed over 3 000 kilometres of spotted hyena spoor.

During this phase of the project my admiration for Hermanus and his stamina grew. The pride he took in his work and the cheerful manner with which he went about it were an example to anyone and made it a real pleasure to work with him. He took it as a personal failing if he could not catch up with the group of hyenas we were tracking and rarely failed his stringent quality demands. If he did, even though it was through no fault of his own, his ever-present smile would be replaced by a scowl and some choice words would be uttered pertaining to the parentage and origins of the particular group of hyenas we had been following.

From our spoor-tracking endeavours with these three new clans and our direct observations of the Kousaunt Clan, we learnt much about the movements and land tenure system of Kalahari spotted hyenas. Like brown hyenas, the members of a spotted hyena clan share a territory, but these are over three times larger than the territories of their shaggy cousins. As with brown hyenas, there were considerable differences in the territory sizes of different clans. The Seven Pans Clan lived in a huge territory of 1 776 square kilometres, whereas the Kousaunt Clan's territory in 1983 covered only 553 square kilometres. Again like that of brown hyenas, the size of a spotted hyena territory seemed to depend on the distribution of food. The further the hyenas had to walk between their meals, the larger the territories were. On average Seven Pans spotted hyenas travelled almost 80 kilometres between kills or carcasses, whereas Kousaunt hyenas travelled only 40 kilometres. Our game counts confirmed that there were more prey animals in the Kousaunt territory than in the Seven Pans one.

A further point of similarity between the two hyena species was that there was no relationship between the size of a clan and the size of its territory – larger clans did not necessarily have larger territories than smaller clans. I have shown that brown hyena clan sizes were fixed by the richness of the food patches in the territory. The Kwang Clan grew when wildebeest were dying in the drought as these were rich food patches. The single female D'Urbyl and her cubs lived in a territory in which most food consisted of small scavenged pieces.

This relationship was not found in spotted hyenas. Their clans did not vary much in size and seemed to be fixed at about 10 members, excluding cubs. This is because the richness of their food patches did not vary with time or in different territories – gemsbok and wildebeest calves and adults were eaten in more or less the same proportions in all clans. There were exceptions to this rule, as there were some very small clans, such as the Urikaruus Clan of four, which I discussed in Chapter 2, and the Kaspersdraai Clan of two adults already mentioned.

I think the reason for these small clans was that a catastrophe had recently hit them, a likely candidate being rabies. My observations of the Urikaruus Clan and later observations of the Kousaunt Clan by

Mike Knight showed that rabies is certainly present amongst Kalahari spotted hyenas and that it can have devastating effects. Once a spotted hyena clan has become severely reduced, for whatever reason, its numbers can normally only be built up again through the remaining females breeding. This is exactly what happened in the Kaspersdraai Clan. When we left the Kalahari in 1984 there were seven adults and three cubs in the clan. All except a new immigrant male were descendants of the original female Marlin. Should a clan be totally eliminated, the vacated territory will probably only be filled by a splinter group, breaking off from a well-established clan, such as the Tu-Tu coalition from the Kousaunt Clan. This was probably the origin of the Seven Pans Clan.

Spotted hyenas from neighbouring clans in the Kalahari are even less likely to meet each other than are brown hyena neighbours. We only witnessed one meeting between two clans. This was accompanied by much running to and fro and a great deal of noise, but no physical contact. However, what may have been the most significant inter-clan encounters were unfortunately clouded by incompleteness. We followed the tracks of four members of the Seven Pans Clan to the Kaspersdraai Clan's den in which there were four small cubs. Two days later when we again visited the den, we found that they had moved. We followed the tracks six kilometres to a new den but found that one of the cubs was missing. Could the Seven Pans hyenas have killed the cub? From the Kousaunt Clan a nine-month-old cub was bitten in the foot and shoulder. At the same time this cub's litter mate and a one-year-old cub disappeared. One month later two six-month-old cubs suffered injuries, but survived – these too may have been victims of hyenas from another clan. In Serengeti, Hans Kruuk and, in Kruger, Joh Henschel also recorded instances of strange hyenas killing cubs. Another form of infanticide has been reported by Heribert Hofer and Marion East in which dominant females in a clan have killed small cubs of subordinate females.

Although Kalahari spotted hyena clans do not vary much in size, if you look further afield there is a great deal of variation in clan sizes in different areas. In Tanzania's Ngorongoro Crater Hans Kruuk found that spotted hyenas live in clans of up to 80 members. Obviously there are numerous adults of both sexes in these clans and, like the small Kalahari clans, there is a strict dominance hierarchy within and between the sexes so that even the lowest-ranking female has a higher status than the highest-ranking male. Even more remarkable is the fact that these sizeable clans live in minuscule territories of 30–40 square kilometres. The Crater's rich volcanic soil and rainfall of 750 millimetres per year make it a paradise for wildebeest and zebra, which in turn provide an abundant and rich food supply for spotted hyenas and enable them to live at this exceptionally high density. Here territorial battles between clans are common and in fact most adult spotted hyenas in Ngorongoro die in clan warfare, either over land or over food. Because the territories are so small the boundaries are clearly marked from time to time by groups of hyenas on special boundary patrols. Hans told me that these territorial boundaries are so well demarcated that if a group of hyenas chases a quarry across a boundary into their neighbour's territory, they will abandon the chase on the border. Sometimes hyenas from the next clan will then take up the chase.

On the neighbouring Serengeti plains, made famous by its million or so migratory wildebeest, spotted hyenas are faced with some special problems. For part of the year a clan will enjoy a superabundance of food as the migrants move through its territory. However, once they have gone there is precious little for them to eat. In September 1989 I drove from the Ngorongoro Crater to Seronera, in the middle of the Serengeti National Park, and saw only a handful of wildebeest on a dusty and dry landscape. Except for the open expanse, the plains reminded me of the Kalahari river-beds in the dry season. In March 1992, in the rainy season, I again made this journey. This time I saw easily 250 000 wildebeest on an emerald surface.

Recently Heribert Hofer and Marion East have discovered a fascinating commuting system in spotted hyenas of the Serengeti. Each clan of 50 or more hyenas defends a territory of some 50 square kilometres. In this territory is situated the communal clan den, which can house anything up to 20 litters of cubs at a time. When the migratory herds leave a clan's territory, the clan members are forced to follow them. This may be a distance of 100 kilometres from the den. Adult females with cubs are forced to commute to the hunting grounds, have a good meal, then return home to suckle the cubs – a round trip which may take several days. Once she has fed her cubs the mother must go back for the next meal. It is only because they are so mobile because of an extremely energy-efficient loping gait, that spotted hyenas are able to achieve this. Anyone who says hyenas are ungainly is talking through his hat.

In the Kruger National Park spotted hyenas have long been regarded as mainly scavengers – the prototype hyena. Joh Henschel's studies revealed that Kruger spotteds do scavenge more of their food than their counterparts in the Kalahari and Serengeti: 80 per cent of carcasses fed on were scavenged. However, when needs be they are quite capable of hunting. Joh saw them bringing down prey as large as buffalo and, as I have mentioned, even on one occasion a hippo. In fact, in terms of the amount of meat actually eaten, half was from prey killed by the hyenas.

Most of the scavenging by Kruger hyenas is from lion kills. However, in contrast to the highly active and aggressive interactions between lions and spotted hyenas which I observed in the Kalahari, and which other hyena observers have witnessed in East Africa and Botswana, scavenging from lions by hyenas in Kruger is usually a far more passive activity. Of over 120 carcasses I have watched lions feeding on in Kruger, I have never seen hyenas even attempt to take over the food. If Kruger spotted hyenas find lions feeding on a carcass, they are inclined merely to lie close by and wait for the lions to leave.

Usually not more than three hyenas gather round a lion kill in Kruger – not a large enough gang to take on the cats. The reason for the small number of hyenas which usually gather at lion kills is not that there are few hyenas in Kruger, for the density there is far higher than in the Kalahari. Rather it is due to the more solitary foraging behaviour of hyenas in this region. Additionally, the ratio of lions to spotted hyenas in Kruger is close to one, whereas in East Africa and Botswana hyenas outnumber lions by 2:1 or more – in the Ngorongoro Crater during Hans Kruuk's study in the 1970s the hyena to lion ratio was about 6:1. This means that in Kruger there is relatively more food available to hyenas from lion kills and therefore less incentive for hyenas to chance their limbs against the larger and more powerful predators, and also less incentive to hunt. Every now and then, however, hyena numbers do build up around a lion kill in Kruger. Then the hyenas become more aggressive and daring and give the lions a far better run for their money, even occasionally forcing them off the carcass.

Although we never found any evidence of siblicide – the death of a cub caused by a litter mate and often called the Cain and Abel struggle in eagles – this phenomenon has been recorded in spotted hyenas. Compared with other carnivores, spotted hyena cubs are born precocious with their eyes open and canine teeth fully erupted. From their captive studies Laurence Frank and his team witnessed fighting between litter mates within minutes after birth. This may lead to the death of one of the cubs or at least a dominance hierarchy between the cubs. Field studies by Kay Holekamp and others in the Masai Mara, Kenya, and Serengeti have found that the dominant cub controls access to the mother's milk and, perhaps even more significantly, that single cubs grow faster than twins and have a better chance of surviving. It seems that the high maternal investment by female spotted hyenas, the lack of communal suckling and the substantial benefits of being a single child have been responsible for the evolution of high neonatal aggression leading to siblicide in the spotted hyena.

This brief review of spotted hyena behaviour shows how adaptable and flexible they are under different conditions. What holds true for one area need not hold true for another. Not only does this add to the fascination of studying these animals in different areas, but it is also very important when it comes to making management decisions about hyenas and, indeed, other carnivores. It is essential to know something about the animals in the area where management decisions are being taken and inadvisable to project results from one area to another, particularly if the habitats are different.

In fact, spotted hyenas have the most complex social system of any carnivore, on a par with that of baboons, members of the most intelligent order of mammals, the primates, which includes ourselves.

CHAPTER 12

Noises
and Smells

MOST CARNIVORES that live in groups have well-developed vocal systems for communicating with one another. The spotted hyena is perhaps the classic example of this. Together with Joh Henschel and myself, Gustav Peters, an expert on carnivore vocal systems, has identified 14 different types of vocalisation in the spotted hyena's repertoire, and there may well be more. Because each one grades into the other, this adds greatly to the subtlety of the system, making it one of the most complex vocal systems in mammals. There is still a lot to learn about the language of the spotted hyena.

In contrast we were only able to recognise eight vocalisations in the less social brown hyena. Not only that, spotted hyenas use their voices more often than brown hyenas do and communicate not only over short distances but, particularly with the whoop call, over several kilometres. Brown hyenas have no long-distance call comparable to the whoop call. The only loud noise they make is a yell by the submissive animal in the rare territorial disputes that take place, as we heard the night that Jo-Ro met the stranger at Rooikop Windmill.

The whoop call of the spotted hyena is one of the most evocative and characteristic sounds of the African night, as much a symbol of wilderness as the wolf's howl. It starts as a deep lowing sound, followed by a slow rise in pitch rising to a high: *whooop*. This may be followed by a descending low, which may occasionally rise again. Each whoop lasts a few seconds, and five to nine whoops are usually made during a bout. I paid particular attention to the spotted hyena's whoop call, noting the situations in which they whooped and also who whooped, in an attempt to understand the function of this very conspicuous and important sound. I soon discovered that each hyena has its own voice and that it is possible to identify an individual by its call. Not only are there subtle variations in tone and pitch, but also in the make-up of the different phases. For example, the male Necklace almost always added a descending low followed by a rise – a sort of double whoop – to at least one whoop in a bout.

Occasionally hyenas answer whoops, but most whooping appears spontaneous in that there does not seem to be any stimulus to start whooping – the hyena merely seems to be saying, 'I am here'. The most common reaction to a whoop is to lift the head briefly as if to listen more effectively.

Sometimes whooping plays a more obvious role. Hyenas that have lost contact among themselves, for example after an unsuccessful gemsbok chase, sometimes whoop to relocate one another. The night that Old Flat Ear whooped 17 times at the gemsbok carcass with the single lioness, she surely was calling for reinforcements. On a few occasions when the hyenas I was following were in the border regions of their territory and heard other hyenas whoop, they immediately turned round and moved away from the direction of the whoops, probably to avoid a clash. If they were in the middle of their territory and I played a tape recording of strange hyenas, this would induce an aggressive reaction: the hyenas would often run bristling towards the vehicle.

The main reason that hyenas whoop, therefore, seems to be advertisement. In the case of immigrant males they are probably reinforcing their presence in the clan. In the case of adult females they are proclaiming their territory. Being the intelligent animals they are, they are able to add variation to the function of the call and use it more when necessary to locate each other.

A further variation on the whoop is the fast whoop – a higher-pitched call, with the intervals between each whoop noticeably shorter. This call is heard during encounters with lions and enemy hyenas from neighbouring clans. It is clearly a call to arms and any hyena that hears it responds immediately by running to the action.

Whereas brown hyenas are inferior to spotted hyenas when it comes to sounds, the reverse is true when it comes to the indirect method of communication through scent-marking. Hyenas scent-mark by defecating at particular areas called latrines. Accumulations of their chalky white faeces, which can almost be mistaken for meringues, are commonly seen along roads and other landmarks in areas inhabited by hyenas. In the Kalahari brown hyenas mainly use shepherd's trees as latrine sites and, inexplicably, almost always deposit their faeces on the south side of a tree. Spotted hyenas hardly ever use any kind of tree and their latrines are typically on the side of a track, next to a dried-out rain pool or on top of a low dune.

A second, more exotic manner of scent-marking is by the unique method called pasting – the secretion of substances from the anal gland on to grass stalks. Again both species paste, but if the spotted hyena is the sound expert, the brown hyena is the aroma expert. Pasting is an intriguing piece of behaviour. The brown hyena approaches the grass, lifts a foreleg over the stalk, turning slightly as it does so, thus positioning the stalk under its belly and the base of the grass stalk between its hind legs. With its tail curved over its back and its back legs slightly bent, it then pushes out the anal pouch, which unfolds like a time-lapse film of a flower opening. For several seconds it feels for the grass stalk, until it comes to lie in the groove that runs down the white central area of the anal pouch. Then the hyena moves forward slightly, causing the anal pouch to slide forward along the grass stalk and at the same time retracting it. The first effect of this action is to smear a thick, creamy blob of paste on to the grass stalk. Then, as the pouch continues to retract, a thin smear of black paste is secreted from the dark area of the anal pouch and is left above the white blob. As the hyena continues to walk forward, the rest of the grass stalk passes between its hind legs, springing back to its original position. A thin black smear on the grass stalk and, below it, a large white blob, both about 15 millimetres in length are clearly visible to the trained eye.

Spotted hyenas paste in a similar way, but they have smaller anal glands, secrete only a white paste, and are not as fastidious as brown hyenas in that they frequently paste on several grass stalks at a time. Particularly with spotted hyenas, but also sometimes with brown hyenas, pasting is accompanied by vigorous scratching of the ground with the forefeet. The pastes of the two species smell quite different: the brown hyena's has an odour that reminds me of caper sauce, while that of the spotted hyena has a more pungent smell. We could still detect an odour from the white pastes of both species over 30 days after deposition, whereas the black paste of the brown hyena only holds its smell for a few hours.

You cannot ordinarily follow a brown hyena for more than about half an hour before seeing it paste. On average a brown hyena deposits 2.6 pastes for every kilometre it walks. In spite of this, until Hans Kruuk's first visit in January 1973, we had never discovered a pasting on our spoor-tracking expeditions. Our tracker Houtop had never pointed one out, in spite of my emphasising many times that I wanted him to disclose everything he could from the spoor. One of the first things Hans asked me when we met was whether I had seen this behaviour in the brown hyena, as he was familiar with it from his hyena studies in East Africa. On practically the first day out with Hans, while following a spoor with Houtop, Hans's sharp eyes quickly picked out a grass stalk with a pasting on it. We picked the stalk and showed it to Houtop. I asked him if he had seen these marks on a grass stalk, to which he replied that he had often done so. When I asked him why he had not told me, he replied that he didn't think that we would be interested to see where the hyena had wiped its bum. After that he regularly pointed out pastings to us and I am sure that he was even more convinced than ever that we were not of sound mind.

A colleague of Hans's, Martyn Gorman, from the University of Aberdeen in Scotland, who is an unabashed expert on carnivore smells, expressed a desire to become involved with the smells of hyenas, and I was more than happy to co-operate with him. First, I sent a number of pastings from various brown hyenas to Scotland for chemical analysis. Martyn came up with some very exciting results. He showed that the chemical composition of the pastes of various individuals is different, although constant for the individual. This implies that the smells produced by each individual are unique. We then ran some simple experiments in which we placed grass stalks with pastings from known hyenas in the paths of others and noted their reactions. These experiments proved that the hyenas are indeed able to differentiate between the pastings of various animals and react differently to them. They will mark on top of a foreigner's pasting but not on top of one from their own clan. How do they know to whom a scent-mark belongs? When two brown hyenas from the same clan meet, they often present their anal pouches to each other to be sniffed. In this way they are able to familiarise themselves with the smells of their own clan members.

When we looked at where in their territories hyenas scent-mark we did not find, as might be expected, that the territory boundaries were necessarily heavily scent-marked. Rather the territory was peppered with marks. This makes sense in these very large territories, as it would take an inordinate amount of time and energy for a hyena to patrol and mark its boundary regularly. In small territories such as those of spotted hyenas in the Ngorongoro Crater the hyenas do mark their boundaries. In the Kalahari most pastings are located in the centre of a clan territory, with a progressive reduction towards the periphery. This is simply because hyenas spend more time in the centre of their territory than they do near the boundary.

This is a good strategy in achieving a regular distribution of scent-marks throughout a territory. Given that a brown hyena deposits a paste 2.6 times in every kilometre it moves, that it moves on average 31 kilometres every night, and that the smell of the white paste lasts at least 30 days, we calculated that, for example, in the Kwang territory in 1977, when six hyenas inhabited the territory, there were in the region of 15 000 active scent-marks at any given moment. Put another way, no matter where a brown hyena happened to be in the Kwang territory, it was never more than 500 metres away from a scent-mark, and over most of the territory it was in fact less than 250 metres away. Given the wonderful sense of smell of a brown hyena and the strong odour produced by the paste, it is obvious that to a brown hyena the area positively reeks of hyena scent. It would not take long before an intruder realised that it was entering hostile country.

Given this efficient way of advertising their ownership of a territory, it is not surprising that brown hyenas from neighbouring territories rarely meet. One of the most spectacular brown hyena encounters we saw was between the Rooikop Clan male Jo-Ro and a strange male, but we did not know for sure that his rival was from a neighbouring territory. In fact from over 1 500 hours of observations of adult brown hyenas away from their dens we saw them meet up with animals that we knew were neighbours on only 10 occasions. The outcomes of these meetings seemed to depend on the sex of the animals concerned. On six occasions the animals were of the same sex and in all cases they behaved aggressively towards each other, in four of the meetings indulging in vigorous neck-biting. Of four meetings between animals of opposite sex they ignored each other in three and actually came together and sniffed each other once.

But that is not all. The brown short-acting paste secreted by brown hyenas may be useful in passing on information between members of a clan. Because they eat mainly small food items, a hyena is likely to clean up the available food in an area fairly quickly. It would be unproductive for another brown hyena to forage at the same time in the same area. I think it is possible that the short-acting brown paste could carry information as to how long ago a brown hyena passed through the area. Then, should another brown hyena from the clan come into the area on the same night, it could avoid wasting its time searching for food in an area that is likely to be unproductive.

Spotted hyenas paste about 20 times less frequently than do brown hyenas. There are also large differences in the frequencies with which various animals in the clan scent-mark. Immigrant males like Hans deposited four times as many scent-marks as the others. Similar calculations of the distribution of pastings in a spotted hyena territory suggested that there are only about 1 600 active pastings in a territory at any one time. Because their territories are also larger than those of brown hyenas, spotted hyenas are usually one to two kilometres away from a pasting.

Although less intense, scent-marking, both through defecating at latrines and pasting, obviously plays a large role in territorial advertisement, in spotted hyenas – it adds a second string to their bow after whooping. By scent-marking both species would, it seems, rather invest heavily in marking the territory, thus warding off intruders, than risk an outright confrontation with an intruder, which might lead to serious injury or death. However, if pushed, residents will fight to defend their land. Intruders know this and they too would usually not risk a confrontation. Immigrant male spotted hyenas also appear to use scent-marking as a back-up to whooping to help establish themselves in the clan and particularly to imprint themselves on the adult females.

I must stress that because our own sense of smell is so poorly developed, we were really groping blindly in the olfactory night when trying to interpret much of the scent-marking behaviour. In fact, following the animals in a vehicle was often very limiting as we were cut off from the outside environment. On top of this we also had the disadvantage of inferior night vision and our hearing, even with the window open and the engine turned off, was less sensitive than that of the hyenas with their large ears. It is not surprising that for much of the time we watched hyenas, we actually had no clue as to what they were doing and thinking. We really only saw the tip of the iceberg, but even that was wonderfully exciting.

As scientifically interesting as the work with Martyn on scent-marking proved to be, our work received unexpected recognition on TV. Sometime later we received a letter from Martyn in which he wrote: 'We have a satirical series called "Not the 9 o'clock News" which exists to take the proverbial out of our leading public figures. This it does with short sketches and the skilful juxtaposition of various bits of film. Last night who should star but a brown hyena and Gus's hands? The clip started with a shot of a brown hyena pasting and then progressed to a close-up of a pile of hyena faeces, and down came Gus's hands to pick up the bits, break them open to show tsama pips, etc. It then went to a shot of Prince Charles delicately placing some object in his mouth, with great and obvious relish. Gus, you are famous at last!' This scene was derived from footage shot by the BBC when they made a film on our work.

CHAPTER 13

Hyena Conflict Zones

DURING THE FIRST SIX YEARS that we followed brown hyenas, we hardly ever saw them meet with spotted hyenas. From spoor we once reconstructed an incident where a spotted hyena had apparently chased a brown hyena away from a windmill. On another occasion, spotted hyenas had seemingly kept the male brown hyena Jo-Ro away from Kaspersdraai Windmill the first night we tried to follow a brown hyena with a radio collar. Lastly, a brown hyena cub, which we thought might be D'Urbyl's daughter Bop, was killed by spotted hyenas near Kaspersdraai.

During the latter part of 1978 this situation changed. For the first time spotted hyenas began to appear in the Kwang brown hyena clan's territory. Early one evening in May of that year while out looking for a brown hyena to follow, I came across two spotted hyenas walking down the road just south of Kwang Pan. What a pleasant surprise! It was like bumping into long-lost friends. It was a beautiful Kalahari full-moon night with excellent visibility and so I decided to see if I could stick with the spotted hyenas for a while. They moved out of the river-bed and soon started loping upwind, coming out at a hartebeest carcass. As we arrived we saw two brown hyenas feeding on the carcass. For the next six hours we witnessed a fascinating series of events.

Predictably, as soon as the larger hyenas arrived, the two brown hyenas withdrew. I had expected that this would be the last we would see of them that night, but I was wrong. The two brown hyenas, whom

142

I identified as Chinki and the young male Thunberg, retreated only a short distance. They then proceeded to circle the carcass slowly at distances varying from about 20 metres to as close as 5 metres, stopping every now and then and staring at the spotted hyenas eating their carcass. The spotteds were obviously aware of them but ignored them and continued feeding.

After over an hour and a half of this behaviour, Chinki came right up to the feeding spotted hyenas, her mane raised in typical brown hyena fashion. The spotted hyenas curled their tails and raised their hackles. I was concerned that Chinki's impatience was going to be her downfall. Then they moved towards Chinki, who retreated a few metres before stopping, crouching and giving vent to a very deep short growl – a sound that I had never heard before. Surprisingly this stopped the spotted hyenas in their tracks one or two metres from her. For half a minute the three animals stood still, Chinki growling two more times, and the spotted hyenas looking around in all directions except, it seemed, directly at Chinki. Then they returned to the carcass and Chinki moved away. Five minutes later she returned and the incident was repeated. After this she lay down about 15 metres away.

During the next hour, two more brown hyenas from the Kwang Clan arrived at the carcass. Over the next three hours, on half-a-dozen occasions at least one of the four brown hyenas approached the spotted hyenas. They would then react by moving towards the brown hyena, resulting in the same stop–growl–scan–depart sequence as had occurred earlier. In between these episodes, the brown hyenas would lie down in the vicinity.

Just before sunrise, Brunnea – one of the later arrivals – approached the carcass with her hair raised and snarling quite loudly. This time the spotted hyenas, instead of coming out to challenge her, retreated, thereby giving her access to the carcass. She immediately started eating. Her success was short-lived as one of the spotted hyenas soon came back to the carcass and dragged it away from her.

Three minutes later Brunnea was back at the feeding spotted hyenas again. One of them dragged the carcass away some 10 metres and continued feeding, but the second one moved away. Brunnea continued to approach and soon the second spotted hyena also moved away. This time neither spotted hyena returned and within 15 minutes all four brown hyenas were enjoying the remains of a carcass that had previously belonged to them.

During the next year I witnessed several similar incidents. In all cases the spotted hyenas were clearly the dominant ones. Yet it seemed to me that by persistently disturbing the spotted hyenas in their own low-key manner, the brown hyenas caused them to abandon the food at an earlier stage than they would have if left in peace. This behaviour seemed to have an effect similar to the far more flamboyant and noisy exhibitions made by spotted hyenas towards lions. There was, however, one important difference: spotted hyenas never lost large amounts of food to brown hyenas, whereas lions sometimes did to spotted hyenas.

When I switched my attention to spotted hyenas and my study area from the Kwang region to Kousaunt, I recorded some interesting and important differences in relations between the two hyena species in the two regions. On only a handful of occasions did spotted hyenas steal carcasses from brown hyenas in the Kwang area and 85 per cent of the scavenged carcasses fed on by brown hyenas were not hijacked by spotted hyenas. In the Kousaunt area nearly all the carcasses available to scavengers were eaten by spotted hyenas. Moreover, of the five that brown hyenas found first, all were taken over by spotted hyenas. In these cases too, the brown hyenas did not wait around and harass the spotteds as they did in the Kwang area. This was simply a result of numbers. In the Kwang area there were usually only one or two spotteds at the carcass and as many as eight brown hyenas, whereas around Kousaunt I never saw more than one brown hyena at a carcass, whereas sometimes there were as many as 12 spotteds.

The difference in numbers of spotted hyenas in the two areas had other consequences. When a group of five or so spotted hyenas from the Kousaunt Clan went on an extensive foraging trip to the south into the Kwang area, they usually fed rapidly on the kill they made and then returned to their den, leaving behind some nice pickings for the brown hyenas. Kills made around Kousaunt, on the other hand, were far more likely to be completely eaten as they attracted more spotted hyenas.

The two hyena species also occasionally crossed paths away from food. Here again the behaviour of the animals was surprising. One night I followed seven members of the Kousaunt spotted hyena clan to Kwang Windmill at which a brown hyena was drinking. They immediately chased it and it ran away. After a good 500 metres, the brown hyena stopped, backed up against a large camelthorn tree and faced the spotted hyenas with its hair raised, ears back, mouth open and snarling. The spotted hyenas stopped about five metres away and looked off in different directions, as if they did not see the brown hyena. After a short while the brown hyena moved away slowly, breaking into a run. Immediately it did so, the spotteds chased it again. This time it could not find a tree to protect its hind region and stopped in the open, adopting the same posture as previously. The spotted hyenas surrounded the brown hyena, and it looked to be in a very dangerous situation. However, the spotted hyenas appeared reluctant to attack it and merely danced around their captive, darting in once or twice to nip it in the back or even merely to sniff it. Then suddenly the spotted hyenas retreated a little and again stood around looking off in every direction except at the brown hyena. The brown hyena slowly and carefully backed up to a nearby bush and after a minute or two slunk off, with one or two of the spotted hyenas watching it go but without showing any further interest in it. Once again the spotted hyenas were obviously in charge of the situation and surely could have torn the brown hyena apart. Yet they showed more curiosity towards it than aggression.

Sometimes when they meet, a strange kind of mutual attraction takes place. This seems to occur when one or two spotted hyenas come across a brown one. Then the brown hyena might actually approach the larger species, as if inviting attack. The two species circle each other at close quarters, the brown hyena displaying its elevated mane, but they do not come into physical contact. A similar kind of mutual attraction was also noticed by Hans Kruuk in the Serengeti between spotted hyenas and the smaller striped hyena.

However, at other times spotted hyenas do attack brown hyenas. One striking incident occurred when Olivia and Goldie came bounding up to a brown hyena. Olivia immediately attacked and grabbed the unfortunate brownie around the jowls, pulling it roughly. The brown hyena gave vent to a very loud and deep growl and Goldie darted around the struggling pair like a referee at an all-in wrestling match. After what must have seemed an age to the brown hyena, but was in reality not more than about 30 seconds, the brown hyena managed to shake itself free and beat a hasty retreat. The two spotteds lunged half-heartedly after it, then let it go. It is possible that these attacks sometimes escalate into uninhibited aggression on the part of the spotted hyenas and may lead to serious injury or even the death of the unfortunate brown hyena.

Although on occasions they are brave enough to stand and face one or two spotted hyenas, they show no such bravery in the face of the whoop call of a spotted hyena. The first time I saw this was when Normali and her grown-up son Shimi were feeding peacefully on a carcass. A spotted hyena came running up to the carcass, at which Shimi ran off, but Normali stood her ground. The spotted hyena stopped in its tracks as if taken aback. It then walked to the other end of the carcass and started eating, while Normali continued feeding. The party did not last long, but it was the spotted's nerve that cracked first and it soon moved away. Twenty minutes later Normali was still feeding on the carcass when a spotted hyena whooped close by. Amazingly, she immediately ran away in the opposite direction.

I tested this reaction of brown hyenas to the spotted hyena's whoop by playing taped recordings of a whoop to peacefully resting brown hyenas. In all six trials the brown hyena reacted. In the first Normali was the 'guinea pig'. Immediately she stood up and ran off in the opposite direction, only stopping after more than a kilometre. On other occasions the guinea pig's reaction was less extreme and it merely walked off some 50 metres before lying down again. These variable reactions are inexplicable but may have something to do with the individual brown hyena's past experience of the jaws of spotted hyenas.

Spotted hyenas also raid brown hyena dens, possibly attracted to them by the smell of the food remains that so often litter them. One memorable occasion again involved Olivia and Goldie. I was following them for the second consecutive night on a hunting expedition that had taken them into the far south of the clan territory, between Kwang and Cubitje Quap. They left the river-bed at the junction of my track to the Kwang brown hyena dens. As they reached the peak of the high dune that leads away from the river-bed, they suddenly broke into a lope, their noses going up as they obviously caught the scent of something. They were really putting on a spurt and I had to concentrate to keep contact with them. After about two to three kilometres they came out at the Kwang den. Three six-month-old cubs of Chinki's were lying on the mound in front of the den and immediately scuttled underground. The spotteds darted into the den mouth after them but were unable to reach the cubs in the narrow tunnels – good proof of the importance of dens for the welfare of cubs.

They soon gave up trying to get into the den and started sniffing around in the vicinity looking for titbits and scraps. While they were thus engaged, a fourth, larger cub of Normali's came to the den. The raiders saw it first and immediately attacked. The cub turned tail, but the spotteds caught up with it within 150 metres. The three hyenas disappeared behind a bush in a cloud of dust and with protest growls and yells coming from the frantic cub. Within seconds the cub emerged from the other side of the bush walking on its carpals or what would be considered its knees, with both spotteds standing over it. Once again it looked as if the spotteds were about to tear the small hyena limb from limb, but once again they held back and were just mouthing at their victim. Suddenly the brown hyena rolled free and darted into a nearby hole, unharmed. The spotteds returned to the den, sniffed around a bit more, chewed on a few small bones, four times went into the den mouth and, finally, 12 minutes after their arrival, moved off, leaving behind, I am sure, four rather shaken but wiser brown hyena cubs.

Overall, therefore, spotted hyenas are unquestionably dominant over their smaller cousins. They deprive them of some food, often harass them and even occasionally kill them. This relationship is important and may be partly responsible for the differences in distribution of the two species in southern Africa. Even in the Kalahari, if I compare the number of brown hyenas I saw when driving around in the spotted hyena-poor Kwang area with those I saw in the spotted hyena-rich Kousaunt area, there was a marked difference: I saw brown hyenas four times more frequently in the Kwang area. It is tempting to conclude that brown hyenas were avoiding the Kousaunt area and the most likely reason for this was the relatively higher concentration of spotted hyenas.

Several other studies have shown that where two closely related carnivores are sympatric – that is, inhabit the same area – the smaller one tends to be more common in those parts of their overlapping range that are not well frequented by the larger one. In the Namib Desert brown hyenas occur along the coast but are absent from the hinterland, which is inhabited by spotted hyenas. In the Serengeti striped hyenas are more common in the less productive areas which tend to be avoided by spotted hyenas. Similar results have been found from studies of tigers and leopards in Nepal, wolves and coyotes in North America, and lions and wild dogs in Africa.

In the Kalahari brown hyenas are able to survive on the many widely scattered small food items strewn across the desert dunes. I calculated that there were over 200 brown hyenas in the South African side of the park. In contrast, spotted hyenas are dependent on an irregular supply of larger prey. Because of the nomadic habits of the antelope in the Kalahari, all areas are devoid of prey for extended periods. The area can support just a small population of spotted hyenas and there were only about 90 in the South African part of the park at the end of my study in 1984. Even if, as I have suggested, rabies had had some effect on the population, this was small. Perhaps at full carrying capacity the South African side could hold 120 spotted hyenas.

Because the southern Kalahari cannot support a large spotted hyena population, the influence of spotted hyenas on brown hyenas is small. At higher spotted hyena densities this influence may well be greater. This is probably the reason for brown hyenas being so rare in the Kruger National Park. The 2 000 or so spotted hyenas that inhabit Kruger provide too much competition for brown hyenas. There is not at present a breeding population of brown hyenas in Kruger. The handful of brown hyenas that have been seen in the park over the last ten years are undoubtedly vagrants from neighbouring areas where brown hyenas are not uncommon, while spotted hyenas have been eradicated.

It was not always so. As I have already mentioned, the first warden of Kruger, Colonel James Stevenson-Hamilton, recorded brown hyenas denning along the Sabie River in the early years of the park's history. Although we don't have numbers it is almost certain that spotted hyenas are far more numerous in Kruger today than they were then. This is not only because of the reduction in poaching, but also because of the proliferation of waterholes and dams put into the park during its development. The water-for-game programme certainly succeeded in increasing certain game species numbers and, as a result, certain predator numbers as well. The swelling of spotted hyena numbers seems to have backfired in the case of the brown hyena, which has been outcompeted by the larger species. This is the only case of a species going extinct as a breeding species in Kruger since its inception, and it is unfortunate that the probable reason for this was the management strategy employed. Today the question of water provision is a hotly debated and intensely researched question in Kruger because of a number of unforeseen consequences it has had for the ecosystem, and the policy has been revised. Interestingly enough, in the absence of brown hyenas, Kruger spotted hyenas have widened their diets to include many more small food items than their counterparts in the Kalahari.

This story underlines the complexity of managing ecosystems. It is so difficult to know when to intervene and, if so, how. The bottom line is that we should manage our parks with all their facets (brown hyenas and spotted hyenas, not just lots of one) and fluxes (numbers going up and down according to variables such as rainfall). We must recognise that ecosystems are dynamic and given to change. As far as possible we should allow these processes to take place. In the great wildlife management debate of laissez-faire versus intervention, I fall in the former camp.

PART 2

Kalahari Days
Margie Mills

CHAPTER 1

The
Early Years

GUS AND I 'officially' met in 1970 when I invited him to my residence
dance. I hate to admit it but after that first night I knew he was the one
for me, and luckily the feelings were mutual. In 1971 I remained at
UCT to finish off my BSc degree in Zoology, and he went to Pretoria
to do Honours in Wildlife Management. Our romance survived the
time and distance and we saw each other as often as possible. By the
middle of that year Gus was busy thinking of his (and my) future, and
wrote to tell me that there was a possibility of starting a two-year
MSc project on brown hyenas in the Kalahari the following year. His
professor said that he wanted someone 'good' to do this project, as it
was going to be difficult and challenging.

I think in most people's minds the Kalahari arouses feelings of
romantic fascination, but at the same time very few people would ever
want to live there. I must admit I knew nothing about the Kalahari
other than it was the home of the Bushmen (San), it was dry and that
my eldest brother, Rob, had been there on a school trip and had had a
wonderful time. As far as brown hyenas were concerned, I don't think
I had ever heard of them, only the more common spotted hyena. But
being twenty, madly in love and enthusiastic about the outdoors, I
was very excited about the thought of this vastly different lifestyle and
had no doubts that I could and would cope. My parents, Harry and
D'Urban Davies, thought that it would be a wonderful experience and
were very encouraging, my mother wistfully saying 'I think I could

live in the Kalahari for two years.' Little did she know that over time she would spend a total of a year visiting us in our somewhat extended stay of 12 years.

Gus was obviously a bit apprehensive about taking me into the bush, for in one of his almost daily letters written while we were apart he wrote: 'I want you, Margs, to think about a couple of things: you might be living in a one-roomed hut next year, we will have very little money, there might be very few people around, you might well be left alone for fairly long periods of time – life could be a bit difficult. Living in the bush for women can be very tricky at times and I want you to really think about this, because it won't be for three months or even three years, but probably for 30 years. Things might not be as bad as this especially after a few years and there will always be clean air to breathe and beautiful things to see, and possibly one day a nice house to live in with a few people around. Margs, I don't want to go into the bush next year alone and I really want to go into the bush.'

He spoke to his parents of his concerns about taking me, a young, fairly wild 'party lover', into the middle of nowhere, but happily wrote, 'My Mum intimated that she thought you were a person of very strong character and could adjust to a different way of life, so you had better watch out 'cause not only do I want to marry you but my folks are keen on the idea too!'

Gus's mother tells the story of when it was time for them to leave the Kruger Park after their first visit when Gus was eight years old, he started to cry. When asked what was troubling him, he replied, 'I don't want to go. This is where I want to live'. So from that tender age his passion for the bush blossomed and grew stronger with every passing year. When he wrote to me saying, 'however much we love each other, and that is unquestionable, you must be able to feel for my work and enjoy it, because this is part of me. We are on the threshold of the most exciting and challenging time of our lives and we've got to work hard together – we're about to really start living and to learn all about life together. This is the big thing that everything is "together" and this is what is so beautiful.' I fully understood what he meant and where he was coming from. So what could I do but agree to marry him. I did have two preconditions though. One was that we buy a potter's wheel that we would take with us so that I could 'throw pots' to my heart's content, and the other was that we honeymoon in Hawaii, neither of which happened.

I started off 1972 at home in Rhodesia with the not very pleasant thought of having to have both my knees operated on. My kneecaps were disintegrating, apparently as a result of all the sport I had played in my youth on hard African turf, and the disintegrated bits were causing friction in the joints. This resulted in a lot of swelling and pain, so much so that I found it very difficult to walk. This had to be fixed before I ventured into the Kalahari. My doctor would not operate on both knees at once, which I thought would have been a good idea, getting it over with in one fell swoop. So I had the one done and six weeks later the other. Gus kept me busy by sending me all sorts of wildlife books and scientific papers to read to prepare me for the future. 'I am glad you enjoyed the book "Innocent Killers" so much – if it didn't excite you then I think you would be the wrong girl for the job you have signed on for!' This book was about spotted hyenas, wild dogs and jackals in the Serengeti. I also passed the time by making my four bridesmaids' and one flower girl's dresses and my going-away outfit.

In the meantime Gus was in Johannesburg trying to get sponsorship for the brown hyena project, then purchasing and licensing the vehicles. It was an infuriating time of empty promises, people taking so long to get back to him, and many other frustrations – so things haven't really changed. I very quickly learnt that Gus was absolutely determined to get this project off the ground and in the end his perseverance paid off. He approached the South African Nature Foundation (now known as WWF-SA) for R13 900. This was for the Land Rover (R2 900), caravan (R2 000), salary and living expenses for two (R3 000 per annum), and petrol, maintenance, etc. (R1 500 per annum). In the end the Foundation donated most of the money and the Wildlife Society of Southern Africa added R2 000, Coca-Cola R700 for a canvas canopy for the Land Rover and Pretoria University R500 for the radio collars that we needed for the hyenas.

Our home was to be a second-hand three-metre-long caravan, which consisted of a three-quarter-sized bed on the one side, a single bed on the other with a table that could open up next to it, a tiny hanging cupboard, a small sink, a two-plate gas stove and a minute fridge with an ice-box compartment that took two ice trays. The caravan also had a side tent, but to give us more space Gus, sponsored by his parents, bought a second-hand Milano Senior 7x4-metre tent sold by a lady who had used it once and then decided that she didn't like camping. New it was R270, but we got it for the bargain price of R150.

Just before we got married and while I was laid low with my knees, Gus went on a reconnaissance trip to the Kalahari. After five days he wrote, 'Living in a caravan and tent is, I can see, going to be a bit tricky and the main thing is not to let it get untidy. Dust is going to be another problem. If we have to sleep in the day we won't be able to do so in the caravan, especially in summer, but the camps are very quiet during the day so we should be able to sleep outside quite easily.' He said that he was 'as happy as a pig in a bucket of apples'.

Undaunted by all that he had been telling me, I tied the knot with Gus on 13 May 1972 in the Anglican cathedral in Harare with the reception in the beautiful garden of our family home. My mother worked hard for months getting it into perfect shape for the occasion. After a week's honeymoon to the Comoros Islands we set forth on our exciting adventure.

Most of our wedding presents were stored in Gus's parents' garage in Johannesburg as those that we had had the chance to view were not suitable for camping in the Kalahari. Two years later when we moved into a brick house, it was like getting married all over again. We had forgotten about most of what we had been given, so it was very exciting opening everything again. I remembered with excitement a rotisserie that we had been given by one of our bridesmaids and couldn't wait to use it. When we eventually unpacked it and I had the chicken ready to cook, I turned it on and was then plunged into darkness, along with the whole camp. It had tripped the camp's power. What a disappointment! The generator was not powerful enough to take any of our electrical wedding presents – kettle, toaster, sandwich-maker – so they had to be packed away again; little did we know that we would not be using them for another 10 years.

Knowing that the Kalahari Gemsbok Park's three camps had only small shops with a few basic non-perishable goods, we did a fairly big shop in Johannesburg before we left. During the next 12 years we were to become proficient at buying provisions in bulk as our visits to town were few and far between. We very seldom went into Upington solely for a shopping trip. Gus was required at intervals to attend various conferences and workshops, usually in Pretoria or Johannesburg, and we always planned our restocking around these trips. On average we went to 'town' two to three times a year. Our record shop was 14 trolleys in Pick 'n Pay.

Arriving at Twee Rivieren, the park's headquarters, I was introduced to the warden and his wife, Stoffel and Judith le Riche, and we then headed for Nossob Camp, 160 kilometres to the north. On the journey I practised saying '*Aangename kennis*' (Pleased to meet you) so that I could greet our new neighbours in their home language – the Kalahari was and still is very Afrikaans – you can't even get an English newspaper in Upington. The people in charge of Nossob Camp were the ranger Elias le Riche, Stoffel's younger brother, and Doempie his wife, who was the camp manager. The previous warden of the park had been Joep le Riche, the father of the le Riche brothers, who were born in the park, and the first ranger (in 1931) was Joep's eldest brother Johannes, so it was a real family affair.

About halfway through the journey we spotted a pride of lions lying on a plain of short bushman's grass. When they put their heads down, they were almost invisible even though the grass was only about 30 centimetres high. Knowing that Gus had an aversion to firearms and that he had made it clear that he would not carry anything when walking in the bush as it would probably be more dangerous for him, a few alarm bells rang. In the beginning, when he did go off wandering in the veld, I made him carry his pocket knife with an eight-centimetre-long blade. I liked to think that if necessary he could protect himself against any unwanted aggressors. Actually I soon learnt that the best way to keep out of trouble is to stay vigilant. The Kalahari is generally such an open environment and there are no elephant, buffalo and rhinos. In addition, lions and leopards are fairly thinly spread and they usually get out of the way when they see you coming.

What excitement I felt as a young bride of 21, being carried over the threshold into my first home as Mrs Mills. Gus had positioned our caravan and tents under the largest camelthorn tree in the camping ground to get the maximum shade in the summer. Within a few weeks we had it all looking very homely and comfortable. I pinned colourful dish cloths on to the walls and made vases of paper flowers to liven up the place, covering storage trunks to convert them into seats. The large tent had a partition at the one end, separating the main part of the tent, which was our sitting or living room, from two small rooms that we used as storerooms. My efforts to keep our house clean rapidly dwindled in intensity when I saw that with the slightest wind, newly dusted surfaces could be used as drawing boards. It was very frustrating, a battle which I soon realised that I couldn't win. Gus's suggestion that we would have to keep our little home tidy was something we always strived for, because if we didn't the situation rapidly became chaotic.

In my first letter written to our parents from Nossob I wrote, 'Life is really tremendous here and I am loving every minute of it.' This wasn't always true, for it wasn't long before we experienced our first Kalahari dust storm. Gus had gone out early to track and I was alone, having decided to do some home chores. I was puzzled by a strange noise. Looking towards where it was coming from, I saw that the sky was dark and blanketed with dust. What I was hearing was the roaring wind, and before I knew it and before I had had a chance to close up the tent properly, I was enveloped by it, a suffocating atmosphere that lasted for at least half an hour.

Strangely a violent thunderstorm followed, and that is when we learnt that our tent was not waterproof and could not cope with strong winds. It got torn in places. Elias advised us to move our set-up out from under the camelthorn tree as they are often struck by lightning. The show of lightning and thunder that we had just experienced made us realise that we must either move or put up some kind of lightning conductor. The thought of coping with full sun convinced us to follow the second option, and soon we had a metal rod projecting above the tree and going into the ground. .

The Kalahari is also a place of whirlwinds, and the first one we experienced in camp was enormously powerful. Once again I heard the wind coming, dashed outside to rescue the washing on the line, but was too late. All I could do was to grab onto the caravan to prevent myself from being blown away. After it had passed I set about the very prickly task of collecting my now very dirty washing that was adorning various thorny camelthorn trees. I found my washing bucket in the dry river-bed 400 metres away. Those few seconds undid hours of dusting and the tent looked as if a bomb had hit it.

I was very keen to become a good cook, for I was sure the best way to my man's heart was through his stomach. Armed with my first cookery book I started out very enthusiastically but that soon dissipated when I found that I usually didn't have half the ingredients required, and that the closest fresh tomato was 420 kilometres away on a roller-coaster dirt road. Having the tiniest fridge and no freezer facilities, everything had to come out of tins apart from a range of dehydrated vegetables, and there are only so many ways that you can make tinned food look and taste good. I am afraid this started my life-long dislike of cooking. There is also very little you can do in a pot on a gas cooker, so I was very pleased to find 'the black box' on my next trip to town a few months later, a bottomless box about 40 by 40 by 40 centimetres which you placed over a gas burner and which became an oven. Because the gas flame was so close to the oven rack, a major problem was burnt bottoms when I was baking bread or cakes, but at least it was an aid to increasing my menu repertoire. I made bread usually twice a week and happily found a no-knead recipe for a wholewheat loaf that we really enjoyed. Thank goodness Gus is extremely easy when it comes to eating and never complained about my cooking, always appreciating whatever I attempted and making light of my disasters. The le Riches were always very kind in sharing any fresh produce they received, and they also had chickens and sheep, so occasionally we were able to buy eggs and meat.

The Kalahari, being a semi-desert, has extremes in temperatures. Living in a caravan and tent, one really felt the elements. Arriving in the Kalahari in May, the first winter rapidly approached and we soon realised that we were not well prepared for it. An SOS to Gus's mum in Johannesburg resulted in the arrival of a huge box full of foot warmers, winter sheets, hot water bottles, scarves, gloves, long johns, vests and beanies. We had many laughs about the way we looked when we had all our winter gear on, as well as blankets wrapped around us. We could hardly move for all that we wore. When we worked at night following hyenas, the heat from the car engine warming up the inside of the cab was wonderful in the winter but hellish in the summer.

Summer was another matter, and given the nature of the work we were doing, we often had to sleep in the day. This was a huge problem for me, especially when we slept on the back of the vehicle. The brown hyena we had followed all night would eventually find itself the right sleeping spot between 7 and 8 am and we would move off a short distance to the shadiest tree in sight. By this time we were usually exhausted, having been awake the whole night, and went to sleep quickly, but when the temperature started climbing I would inevitably wake up within the hour, flies pestering me, and that would be the end of my sleep. Gus trained himself to sleep in all circumstances, which was vital when one had to keep going night after night. I would try to lie still so as not to disturb him, keeping myself occupied by reading or by thinking about all sorts of unimportant things, but I still found it quite challenging.

If we ended our night's observations within a reasonable distance from camp (less than an hour's drive) we would return home to sleep there. The caravan was often impossibly hot, and even using the soon-developed technique of soaking a towel in water, gently squeezing it out, and then lying on it, keeping a jug of water next to the bed so we could wet the towels again when they dried, we still struggled to sleep. When we became desperate we would move to the camp-ground ablutions with our mattresses, Gus into the Gents, and I into the Ladies, and hope that we would not be disturbed by the cleaners or tourists. Inside the ablutions it was darker than in the caravan and it retained the coolness of the nights longer than our caravan did, which also made it more conducive to sleep. Because of our nocturnal activities we usually ate twice a day. Our main meal would be when we woke up, which was usually before midday. I made sandwiches for the night and whenever possible included a few snacks and treats to spice up the night: the thought of them would help us carry on through the boring, inactive times.

In December 1972 we went away for Christmas to our families, as well as to escape the oppressive heat for a while. When we returned in January we heard that 38 millimetres of rain had fallen in less than half an hour. With great trepidation we unzipped our home and were horrified to find that the water had poured in, particularly where the two tents joined. The water level must have been pretty high because our one and only decent piece of furniture, a cupboard, which was standing in the middle of the tent and which housed most of Gus's precious books and scientific papers, had soaked up the water. Most were damaged. I tried with limited success to steam apart the pages of the really special books and we salvaged what we could. I then discovered that our 'waterproof' trunks, which we used as seats, and which housed all my winter clothes and sewing materials, were full of every kind of mould imaginable, and the contents were mostly permanently stained and unusable.

The mail was our most important lifeline with the outside world. There was no phone at Nossob Camp, the nearest one being at Twee Rivieren, 160 kilometres away. A South African Railways truck used to visit Twee Rivieren from Upington once a week delivering goods and post. We then had to wait at Nossob for someone, either a Parks Board employee or a tourist willing to play postman. Our parents were very good at responding to our weekly letters (although they weren't always posted weekly because of lack of postmen), and we eagerly awaited their replies. We could only pick up short-wave radio programmes and the reception was often not very good. LM Radio, the BBC World Service and Springbok Radio were the stations that we most often tuned into, although Springbok Radio stopped broadcasting on short wave in 1979. I really missed being able to listen to some favourite programmes which I relied on to help me feel not too lonely when I was in camp on my own.

Generally, the Kalahari, because of its low human population density, clean air and dry summers, is a healthy place to live in, although apparently the first warden and his assistant died of malaria in the early 1930s. Only once did we make a special trip to Upington when we were concerned that there was blood in our stools. The doctor asked whether we had been eating anything colourful. It turned out that bottled beetroot, a regular part of our diet, was the culprit.

Occasionally we witnessed exciting happenings right from our tent. Early one morning our mentor, Hans Kruuk, who was staying with us, called us to come quickly. He had just seen a leopard walk past the camp fence carrying a jackal in his mouth. We dashed out of camp in the vehicle and soon came across a rather bedraggled-looking jackal lying under a blackthorn bush. We thought it must be dead, but as we approached closer it jumped up and ran off. The leopard had caught a springbok in the early hours of the morning, and the jackal was obviously a little bit too cheeky. The jackal had obviously fooled the leopard into thinking that it was dead – behaviour reminiscent of the brown hyaena and lion episode at Cubitje Quap mentioned earlier.

Having vowed that I would never smoke, I stupidly started doing so during exams at the end of my first university year, thinking that it calmed me down. When we got married Gus said to me that if I had to smoke, then I must, like him, smoke cigars or a pipe, because it was better for me. He was an extremely disciplined smoker, often limiting himself to only one smoke a day. So that is just what I did. Most people thought it was rather 'unladylike' to see a female puffing away on a pipe, but my Dad had had a continental girlfriend in his youth who did the same, so he didn't think anything of it. When we moved to Kruger in 1984 several people told me they had heard of a crazy woman from the Kalahari who smoked cigars *and* a pipe. I know that many people thought that I was somewhat eccentric for doing so and I suppose I must accept that it was rather unusual. Gus has a 'blackmail' slide of me sitting in the tent, hair in curlers, smoking a pipe.

When sitting at a brown hyena den, which we did frequently at the beginning of the study, we would take turns to be on watch for three-to four-hour stints. The watcher would sit in the front of the vehicle and the sleeper would lie at the back. Smoking was one of the things that helped keep us awake when nothing was happening, which was most of the time. We would work out a smoking plan, allowing ourselves a few puffs on a cigar every couple of hours, stubbing it out until the next time slot. Interspersed with the smoking were the breaks for a drink and something to eat, but even with these incentives we found it very difficult to keep awake, especially between 2 and 4 am. I was quite a night owl, but having to sit still and quietly in the dark was different. I crocheted countless metres of lace, feeling my way in the dark, often undoing the night's handiwork the following morning because of unacceptable mistakes. We could not even listen to the radio because we didn't want to disturb the hyenas or miss out on any night sounds.

When we started following animals at night, we had to learn several new techniques. I struggled to hold the rather cumbersome radio antenna and cling to the back of the bouncing, moving vehicle at the same time. We would take turns, but I also found it difficult to drive the vehicle without using its lights, and the bushes often seemed to jump out at me, appearing gigantic in the night's darkness. The vehicle did not have power steering, and I quickly learnt that I mustn't grip the steering wheel but hold it lightly on the outside, making sure I had no thumbs on the inside. It would sometimes spin violently when we hit a hidden hole and the cross-member would strike my thumbs if they were in the way. We used a hand-held spotlight to see where we were going and maintain contact with the animals.

I got very experienced at pouring cups of coffee when we were traversing the dunes, and felt I would have coped admirably as an air hostess during turbulent flights. We had to do this because the brown hyenas never seemed to stop once they started foraging in the evening. We had to keep our eyes peeled because they seldom went in a straight line, circling and searching, and we couldn't risk losing sight of them by stopping. We used to giggle when drinking our coffee out of pink and blue baby cups, but that was all that was available on the market at the time.

We also got very proficient at peeing from a moving vehicle. Gus had a somewhat different technique from mine. I had to open the door and hang over the edge, gripping onto the body of the vehicle with one hand and the open door with the other, feet half on the floor and half out of the car – a good exercise in building up the biceps. Occasionally Gus got a thrill by driving, on purpose, for a particularly large and thorny bush when I was busy going about my business. Of course he always pleaded innocence.

One thing that I really struggled with was all the sitting while waiting for or watching hyenas. Having always been a very active, sporty person, I quickly got uncomfortable sitting still in one position, and found that the only possible way to relieve the pressure on my buttocks was to do head stands on the front seat.

I suppose few marriages are conflict-free, but I learnt very quickly that it wasn't a good thing to have a tiff with my husband. I had no means of escape: no phone to chat to Mum or friends, no car to get away in, nowhere to go and nothing different to do, like window-shopping or buying myself a new outfit to make me feel better. I did once seriously think of climbing the nearest windmill and waiting for a car to pass, but at that time tourists were few and far between and I would have had a long wait. So we realised that we had to make the most of the situation and learnt to sort out our differences quickly and amicably. Now that I think back, it was a really good way to start a marriage, away from everyone and everything, because we could never run away from each other or our problems.

One of the funniest incidents we had while living in the tent and caravan set-up was when Jan Nel from Pretoria University, who was working on rodents in the Kalahari, came for tea one afternoon with his assistant. It started to rain very gently. After a short while water started seeping in through a tear in the ground sheet and half an hour later we found ourselves sitting with our feet in 15 centimetres of water. Moving our feet created water currents which swilled around us, and to my horror I saw a white object come floating out from behind a cupboard. It was the enamel bed pot which had been tucked away out of sight. What an acute embarrassment! Actually it turned out to be very useful as a floating ashtray circulating between the four of us. At least there was something to laugh about. I really felt like crying at all the damage that the water was causing.

During this storm we learnt that our abode was positioned at the lowest point of the camping ground, so we decided that we had better move if we did not want to be flooded every time it rained. Unfortunately because there were no other large shady trees available, we had to move to an open area on higher ground and contend with the full blazing sun. The year 1974 proved to be very wet and within two weeks we were flooded again in our new position. As there was nowhere higher to move to, we were allowed to deposit all our vulnerable furniture and goods in the camp museum 200 metres away. This left just a table and four metal chairs in the tent, which were easy to dry. It was not an ideal situation having to prepare meals in the tent, with all our food stored in the museum, but at least we got a lot of exercise running backwards and forwards.

All this rain also caused an explosion of flies, which nearly drove me crazy during the day and a myriad of other insects that did the same at night. As we were open to the elements it was impossible to keep them away from us and at times we were ashamed of our language. Summer also meant snakes and scorpions, and I was always on the lookout for signs of them around the tent. The surrounding sand could tell us whether we had any unwanted visitors that had slithered towards the tent rather than away from it.

The two most dangerous snakes that occur in the Kalahari are the yellow cobra and the puff adder. Luckily the cobra is a very shy and non-aggressive snake and is always very keen to get away from humans. If you get bitten, you apparently don't have more than half an hour to get help. The puff adder usually 'puffs' when you are near it. I had several incidents when I could hear the alarm signal but could not see the snake and had to jump backwards hopefully to get out of harm's way. When I eventually spotted it I realised I was close enough for it to have bitten me, but obviously it didn't feel the need to.

Scorpions are particularly visible in summer when a strong wind blows and it is not uncommon to see half a dozen at a time going about their business. The very toxic *Parabuthus grandulatus*, with its small pincers and large poison sac, is common in the area and its sting can be lethal, especially for children. Gus was barefoot once and squashed something underfoot. After apologising, thinking it was a beetle, and lifting his foot, he saw that he had flattened one of the very poisonous species. Luckily its tail had been pointing downwards and went into the sand and not into Gus.

Fortunately we never had to deal with snake bites, but sometimes people in camp received scorpion stings. One night two staff members informed us that a 15-year-old girl had been stung by a scorpion and they thought she was going to die. Gus had to go and reassure her, taking a packet of frozen sausages to put on the bite. We had no frozen ice blocks in the deep freeze at the time. Another time a young girl got stung by a scorpion and her mother insisted that she had to place the now dead scorpion near the sting. I think the patient screamed more because of the sight of the pain inflictor than from the pain itself.

One afternoon during this rainy time we were informed by some tourists that they had seen a dead brown hyena near Kaspersdraai. Even though it was getting late we set off to investigate, taking Houtop, our tracker, with us. After a lot of fruitless searching we decided to cross the river-bed and look on the opposite bank. We didn't get far before we found ourselves stuck axle-deep in clayey mud. By this time it was dark. Gus asked the tracker to help him collect some dead wood to put under the tyres, but Houtop refused, saying it was now dark and too dangerous as there were lions around. Gus decided that he was being needlessly cautious and instructed me to stand on the roof of the vehicle and shine the spotlight for him under the large camelthorn trees about 100 metres away where there should be a good source of dry wood. No amount of pleading from me helped and he strode off into the darkness.

I was far from happy about the situation and must admit that I thought that Houtop would definitely know better than Gus, having grown up in the Kalahari. Shakily I shone the spotlight from my perch, not only on the requested tree but all around to check for any lurking beasties. On one of the searching sweeps Houtop and I spotted some eyes; he quickly informed me that they were lions. I screamed at Gus, imagining having to sit and watch him being devoured by the predators whilst I was stuck helplessly in the mud. I felt sick and decided that I didn't want to be subjected to this sort of thing anymore. Gus rather crossly made a rapid retreat to the car brandishing logs for protection, only to discover that the eyes belonged not to lions but to spotted hyenas. As he was a hyena man they would never do anything to him, so what was all my hysteria about? We tried to jack the car up and force wood under the tyres but the vehicle wouldn't budge.

At this stage we were staying with Elias and Doempie for a few days as our tent had been badly ripped and they had kindly rescued us from all the mess. Luckily Elias had a feeling that something wasn't right and came in search of us. The next day we had to use a tractor to pull the vehicle out. I was not impressed with my husband. The whole episode emphasised the differences between us: me ultra-cautious (wearing braces and a belt, as Gus would say) and Gus very laid back and not worrying about what could or might happen. I have to admit that I have usually worried for nothing.

By January 1974 we had been told that Gus would be employed by the National Parks as the first Kalahari biologist, starting in June. Gus wrote to our parents, 'I must admit I even feel now that it will be very nice to get into a proper house and can only marvel at the way Margie has coped.'

The rain continued and there were huge floods in Upington. The Kuruman River, which had last flowed in 1920, was in flood again, with the school and post office in Van Zyl's Rus under water. Gus's folks had to cancel a trip up to visit us as the only way into the park was now via South West Africa, which was a huge detour. The Auob River was still flowing one metre deep compared with the previous year, when it was only about 10 centimetres deep. By April Mata Mata Camp had run out of fuel and Nossob's stocks were getting very low, so we all had to curtail most of our driving, having no idea when fuel tankers would be able to get through to us again. We went without any post or contact with the outside world for six weeks, which seemed like an eternity at the time.

One of my jobs when we were in camp was to analyse the hyena scats that we had collected in order to assist in working out their feeding habits. It was a painstaking job. After drying out the scats in the sun, I ground them gently in a pestle and mortar, being careful not to crush any insect parts and seeds. We very soon discovered that the hyenas had a very wide diet. The seed remains, mainly from the various melons, reptile scales and bird feathers were easy to extract and identify, but it was the hair analysis that was the complicated part. Because it is a scavenger the scat of a single brown hyena can contain the hair of several different mammal species. From various kills that we came across in our travels through the park, I made a comprehensive collection of the hair of all these animals. Hair had to be collected from all parts of the body – leg hair differed from torso hair, for instance. By early 1974 I had managed to retrieve all the identifiable bits from 383 brown hyena scats and 23 spotted hyena scats, and had only a handful more to go, but of course Gus kept collecting, so it was a never-ending exercise. I wrote, 'We found a new spotted hyena den near Kaspersdraai with lots of faeces. I'm beginning to detest the sight of those white lumps of ****!'

We managed to get out of the park at the end of April and went up to Johannesburg to replenish our food stocks and visit the South African Police forensic department to learn how to identify hairs. By making imprints of the hair in a gelatine solution on a glass slide and then looking at the shapes of the hair scales under a microscope, and also by making cross-sections of hairs using a microtome, one could distinguish between species, although sometimes the differences were very subtle. When we returned to the Kalahari I made imprints and cross-sections of all the known samples of hair we had collected, so that I had a good source of reference when analysing hyena scats. Later on I also spent two weeks at the Transvaal Museum working on hair analysis and claw identification, using their collections as a comparative base. If only we had had the foresight to collect and store hair samples from hyenas for the DNA era. I wonder how much more detailed our findings would have been.

Another of my camp jobs was to plot the movements of hyenas.

When following spoor with a tracker we used a compass for direction, having to move at least 10 metres away from the vehicle as the metal of the car affects the compass readings. At night when following the animals, we would regularly take odometer readings and obtain the direction of travel mainly by using the moon and stars. There was no time to use a compass. Knowing our starting and end points and the distance we had travelled through the night, I would make the movements 'fit'.

When it was cloudy and no stars were visible, we really struggled and had many disagreements about the direction we were travelling in. It was amazing how disorientated we would get at times, especially when the hyenas were moving round in circles and we would be absolutely sure that the moon was on the wrong side of the night sky. How wonderful it would have been to have a GPS and how much more accurate our data would have been. It would also have saved many arguments about the quickest way to get home. Luckily we always knew on which side of the river-bed we were but we did get 'lost' on a few occasions and had to wait for the sun to rise before we knew in which direction to drive to reach camp.

Occasionally I had to stay in the tent on my own for a few days. Gus usually visited professors Eloff and Bothma, who camped up at Dankbaar in the dunes in the north of the park. This was a rustic camp from which they conducted their research on lions and leopards, doing so at least twice a year for a week or two at a time. It was always a solely male affair. I wrote to our parents: 'I haven't been lonely while Gus has been away visiting the Profs, as a very nice couple from Welkom saw I was on my own and asked me to eat with them every evening. We usually chatted until after midnight!' Generally visitors to Nossob were very friendly to us. Our unusually permanent-looking abode under the dominant tree in the camping ground often caught their attention, and they were keen to find out what we were doing there. We made many lasting friends during our camping days. During one particularly bad dust storm when I was on my own and being rocketed from all sides, a tourist from one of the huts braved the storm and called me to come and share the comfort and security of their hut with them. We remained friends for years, often staying with them in Schweizer-Reneke on our journeys to and from the park.

One of the drawbacks of living so far away from anywhere was that we missed out on several special family gatherings, which I found difficult to deal with. I did not get to three of my four brothers' weddings. Two were held overseas but the third was in Zimbabwe and I so badly wanted to be there. Gus said that he would drive me to Upington and I could fly from there, but because of all the rain the road to Upington was in a bad state and it would have taken more than a day to get there. I felt very sad about not being able to be present.

CHAPTER 2

No Longer
Happy Campers

IN JUNE AND JULY 1974 we went to the Kruger Park to try to establish whether brown hyenas were resident there or whether the few sightings that had been recorded in recent years were just of vagrants. Gus was now officially employed by the National Parks Board. Many of the promises made to him about vehicles, accommodation and budgets for this trip were not forthcoming and we found it rather disconcerting on arriving in Kruger to find that hardly anyone knew about our visit. Thanks to two influential friends in Skukuza, Salmon Joubert and Butch Smuts, we were able to accomplish our mission. Having travelled throughout the park we could find no evidence for a resident brown hyena population.

At this time a new house was being built for the le Riches at Nossob, and we could only move into their old house once they had moved into the new one. Gus came back to the Kalahari at the end of July for a meeting with his new bosses, who were visiting the park, while I stayed in Johannesburg in order to buy all that we needed to make our new home habitable.

The outcome of Gus's meeting was a number of decisions and agreements: the marking of antelope species would become his most important project after the hyena work; he must learn to shoot and would be given a rifle (in fact this never happened because of Gus's aversion to firearms); we could have a meat ration of one springbok a month or one gemsbok every three months; we could also keep a few sheep though we declined. The house would be painted white inside (it was then every colour under the sun) although it was not exactly clear if the Parks Board would pay for the entire job. Both the old office and the museum would be Gus's territory.

I left for Nossob the day before my 24th birthday in early August and drove alone from Johannesburg in our Hilux bakkie. After eight hours I got to Kuruman and the next day went on to Twee Rivieren, getting there by midday. I then set off for Nossob Camp and 35 kilometres later noticed a windmill sign 'Gemsbokplain' and with horror realised that I had driven up the Auob river-bed and not the Nossob river-bed as I was supposed to. I really don't know what happened to me – I was day-dreaming about my husband but I didn't think I could make such a bad mistake. I then had to drive all the way back to the confluence of the two rivers just outside Twee Rivieren as I didn't want to risk going over the dune road that joined the two rivers. If I had had car trouble they would never have looked for me there. It was a very lonely birthday and I spent a lot of it crying.

A week later we moved into our first brick home and I wrote to the folks: 'It's absolutely wonderful to be in a house – Gus and I are like two excited children with a new toy. We keep running in and out of each room just to check that it's really there, swinging our arms around really savouring the luxury of so much space. It's real paradise. We've just had a bit of rain and the winds have been howling. I keep reminding myself that I don't have to go and tie things down – wow, I can't believe it.' After the first dust storm I seriously questioned the logic of having air bricks in the outer walls of each room and quickly covered them with paper to stop the sandy dust from pouring into the house.

I got stuck into making all the necessary curtains, now having the luxury of a treadle sewing machine instead of a hand machine. With no electricity during the day it was pointless getting an electric sewing machine. We also tried to make a bit of a garden and very ambitiously transplanted a three-metre-high shepherd's tree which took us two days of backbreaking work to dig out, as well as three small camelthorn trees. Because of the severe dryness at times, all perennial plants in the Kalahari have incredibly deep root systems in order to survive; a 10-centimetre-high plant can have a root system half-a-metre deep. Needless to say, the shepherd's tree didn't survive and only one of the camelthorns made it. As it is an extremely slow-growing tree, we did not gain any shade from it during the ten years that we lived in the house.

Using our new meat ration, a privilege then given to employees living in remote areas, we ambitiously went for a gemsbok, which we soon realised was too much to cope with. We made tons of biltong. Doempie also lent me her electric mincer and sausage-maker and we made metres and metres of sausage, storing it in the camp freezers as we did not yet have a deep freeze of our own. I eventually purchased our own sausage-maker, an old-fashioned hand mincer with a spout attachment over which one fed the sausage skins (sheep intestines, obtained locally, preserved in masses of salt). It was exhausting making 40 kilograms of sausage at a time and was certainly not my favourite pastime. Whenever possible, I made sure that Gus was around when the meat needed mincing and the skins needed stuffing. The end result kept us well fed for months. After our first ration we decided to go for a springbok, which was much easier to cope with, and we became well known for our springbok fillet braais. I marinated the fillets for at least a week in wine and various herbs and spices, which made the meat melt-in-the-mouth delicious. Occasionally we were generously given a sheep by Elias. On Gus's birthday one was led down the hill to our house, bouncing and bleating and adorned with a red ribbon round its neck: Gus's birthday present. I wasn't keen to eat it after that.

At the end of August we were rather sad to see our first home being towed away. A Pretoria student came to fetch the caravan, as he was about to start a project on cheetah in Etosha National Park. We very soon heard that he had complained to the university that it was far too small for him and his wife to live in, and it was quickly replaced by one twice the size. We didn't think to complain. Interestingly he and his wife separated later whereas we are still together after 37 years, so perhaps the smaller caravan was better for bonding!

In September 1974 we spent our first night following spotted hyenas. In my letter home that week I wrote: 'What fantastic creatures spotted hyenas are. We had such a super time with them the last few nights – to see them hunting in the moonlight is unbelievably exciting – sort of sends a shiver down one's spine. We went up to the den at Urikaruus in the Auob River but no one was around so we drove into the river and suddenly spotted Short-tail, her face covered in blood, carrying a hartebeest leg and running back to the den. On arrival she dropped the leg and proceeded to call out little Piggy, her pitch-black cub. Short-tail suckled her for half an hour and then moved off. Piggy found the leg, picked it up (an amazing sight as it was about twice as big as it was) and carried it into the hole. We followed Short-tail until she lay up under a bush and found ourselves the shadiest tree in the area to spend the day.

'That evening when we returned to the den Short-tail was again suckling Piggy. At 10.30 pm two other hyenas appeared, Long-tail and her fairly large cub. Suddenly at 2 am they got up and ran off. We drove after them as fast as we could but lost them. So we quickly got out our 'last resort' – the tape of laughing hyenas which had the desired effect and very soon Long-tail came running towards us, excited about the prospect of some food. I really felt mean for having conned her but we did not do it often. After a few minutes of just waiting around the car she started moving off and led us to Short-tail and her cub. We followed them into the dunes and all of a sudden they started running, we thought after something, but it turned out to be three adult female lions and their five fairly large cubs on a kill. For the following two hours the hyenas kept approaching the lions, all cocky, tails up and looking fantastic, only to be chased off a short distance. I was concerned during the whole episode that the hyena cub would get nabbed by one of the lions but thankfully it stayed in the background when the adults were challenging them. The lions eventually left and the hyenas were quickly onto the scant remains. They then moved back into the river and down to the waterhole, Long-tail and her cub playing fantastic games chasing each other all over the place and running at incredible speeds. They got back to the den at about 5 am and we moved off and slept.

'The following evening we sat at the den until 10 pm when the adults moved off. After a short while they dashed off at high speed and we lost them. Luckily we heard whooping and, following the sound, were thankful to meet up with another four animals that were busy greeting each other with lots of grunts and growls – an amazing ritual. They then moved down the river and suddenly dashed up onto the calcrete on the side of the river where they found the same pride of lions as the previous night. Once again there was a lot of antagonism and they kept challenging each other – an incredibly noisy interaction, especially when one is sitting in the middle of it all. Gus uses the analogy that the lions are like the Mafia as they behave in a smooth and superficially socially acceptable manner, whereas the hyenas are more like the street gangs, noisy and rough.

'Eventually they parted company and we continued down the river with the hyenas to the picnic spot where they raided the rubbish bins, standing on their back legs trying to get at the contents inside – a rather undignified sight we felt.' Thank goodness there are no longer containers at these various spots in the park and visitors have to take all their rubbish with them.

'Back into the river they soon came across four wildebeest. We thought there was going to be some action but the wildebeest stood their ground while the hyenas surrounded them, obviously testing them. If one of them had run, the game would have been on. They then disturbed a herd of springbok that pronked off spectacularly in the moonlight. Shortly afterwards they went off into the dunes, having sensed (smelt?) some gemsbok. One of the clan dashed off towards them and we careered after him, driving like mad to keep up and hoping that we would not land up in a hole or crash into a tree trunk! It was a real adrenaline rush. The gemsbok got away and the hyena suddenly realised that he was on his own with us. After searching rather frenetically for his clan members he then seemed to resign himself to the fact that he was on his own, changed his behaviour completely and started foraging like a brown hyena.'

This was the first of many incredible nights we spent with these amazing animals. The excitement of watching them do what they had been doing for thousands of years while seemingly oblivious of our presence was to us an indescribable privilege.

This was an extremely frustrating time as far as the brown hyena project was concerned. The first two collars that we put on an animal soon stopped working and the third one had a range of only about two kilometres. I felt extremely sorry for Gus. The capture pistol had also packed up and the vehicle was giving trouble. Gus was trying to get new collars to put on the animals. In July he asked a firm in Johannesburg to order them from a company in America, explaining how important it was to get them as quickly as possible, for without them the project was at a standstill. By mid-September we discovered that they had still not been ordered. Gus even talked of calling it a day and just writing up what information he already had for an MSc, but being the determined person that he is, he waited patiently until just before Christmas for the equipment to arrive.

This was one of the major frustrations of living in the Kalahari – trying to get broken equipment mended. It was always a major logistical problem to send things to the city to be fixed, and of course no one was in so much of a hurry as we were. With no phone for us to keep badgering for the repairs to get done swiftly, we often felt it was a case of 'out of sight, out of mind'. At one stage we spent several months without a work vehicle and about six months without a radio receiver. We needed the receiver to pick up signals from the collared brown hyenas. It was always such a mission to catch the animals to put the collars on them, and the thought of not being able to make the most of it when they did have collars on was, at times, almost too much to bear. We always kept ourselves busy with other projects to try to lessen the frustrations. Those who think that our life was an idyll, I can assure that it wasn't, and we felt mighty 'stressed' at times.

Living became easier for us at Nossob Camp when we acquired a paraffin deep freeze. It worked like a bomb in the winter, but in the summer I had to plan very carefully, deciding beforehand what I was going to need out of it for the day. I then kept it open for as short a time as possible, otherwise the food would not stay frozen. Now we could also keep frozen vegetables as well as meat and, with oranges buried in the sand (where they would go hard but keep their juices for ages), we began to have a more balanced diet. It became even better when we could afford a fridge.

We spent our first Christmas at Nossob in the company of friends from university days. It made it easier being away from our family for the first time. When Gus took all the visitors out, I had to stay behind because there wasn't room for me. It was hard to miss out on all the fun and games. This became a source of some resentment for me when we had visitors. Of course I was very keen for them to go out and experience the wonders of hyenas, but it often meant that I had to stay at home on my own.

To say that our lives were unusual is putting it mildly. We spend New Year's Eve of 1974 sitting alone, in the dark, at a brown hyena den, and saw absolutely nothing. It was at times like this that I wondered what I was doing in the Kalahari. Thank goodness the next night was more exciting when we followed Normali. She was chased by a bat-eared fox, which ran circles around her, yapping its shrill alarm call. She then smelt the remains of a hartebeest calf that had obviously been killed by a leopard because it was up a tree. 'Our poor friend was so frustrated,' I wrote; 'she just kept looking up at that delicious chunk of smelly meat and cursed the heavens for not having given her wings! After about half an hour of wondering what to do about the matter, she wandered off a bit and decided to sleep on it. So we moved off and slept ourselves within "beeping" distance so that we could keep checking that she was still there.' Later she showed a lot of interest in a honey badger, which went into a thick bush emitting ferocious noises, a very scary sound in the middle of the night. She then picked up a cached mongoose and took it back to the den, being chased by a lioness on the way. After that adventure-packed night, I realised that we had to take the good with bad, and that the most important virtues to have in this job were patience and perseverance.

In 1975 pipes were laid from Kwang Pan to bring 'fresh' drinkable water to Nossob Camp. Until then the tap water was strongly brackish and caused havoc with the metal pipes, rapidly corroding and blocking them. The only drinking water we had was from a 5 000-litre rain tank attached to the house. We treated every drop like gold, not knowing when the tanks would be replenished from the heavens again. The advent of this new water allowed us to attempt to grow vegetables and make some sort of a garden, although we were not entirely successful. What seedlings didn't get eaten by insects and squirrels were flattened by winds, and the grass lawn (only a few square metres of it) became a food source for the ground squirrels and ants. Over time we did raise a few lettuces, carrots and beetroot. We ate these with pride and relish, but the effort far outweighed the end result and we gave up trying after a while.

When in camp our activities were very much affected by the weather. I never got used to the heat of the summer and found it completely energy-sapping. The heat was relentless, the temperature hovering above 35°C for weeks on end. If I hadn't cleaned the house by 9 am, then it did not get cleaned – it was just too hot to worry about such things. In the evenings when we went to bed, we would open all the windows wide and the following morning by 9 we would close up the house and draw all the curtains, trapping as much of the cool night air as possible. As the camp's generator was never on during the day – only for an hour or two around radio session time in the early morning, and then again in the evening from sundown until 10 pm – we did not have the luxury of fans during the heat of the day.

Winter also had its drawbacks. With temperatures below freezing for nights on end, occasionally the gas froze in the pipes and we could only use the gas stove or water geyser when the sun had warmed up the pipes, which was usually well after 10 am. The coldest we recorded was –9.9°C. We spent a lot of time in the kitchen, using the gas oven as a heater. The houses in the Kalahari were not built with the extreme weather conditions in mind. The dryness of the area was also difficult to deal with, particularly in the winter, and I had to use copious amounts of body cream and Vaseline to prevent myself from cracking.

The rains of 1976 caused havoc in the park as well as in many parts of the country. The approach roads became impassable, and those inside weren't much better, even for four-wheel drive vehicles. By February the park was closed to tourists as the Auob was flowing again and the water had already reached Twee Rivieren. The grass grew so long that it was difficult for us to follow the hyenas and our activities were severely curtailed. In early March 125 millimetres of rain pelted down at Nossob and 200 millimetres at Mata Mata in a single hour. All bookings had to be cancelled until 1 May. A few tourists were allowed in from the South West African side but they had to bring their own fuel with them. In October 1976 tough petrol restrictions were introduced because of the worldwide oil crisis; this certainly had an impact on the number of tourists who were able to get to the park. As we left the park so seldom, it did not affect us too badly, although we had to plan our activities in the park more effectively.

Hans Kruuk came to visit us again in July 1976. It was wonderful to have him as a supervisor and he helped Gus immensely, being so full of good ideas and suggestions. As a scientist Gus often felt isolated, having no one other than his wife to bounce ideas off, so he relished times like this. During Hans's visit we conducted a game count from Nossob Camp up to Union's End, then down the South West African border to Mata Mata, going on to the Namib Research Station for a few days. We were planning to spend the first night at Mata Mata, but about 20 kilometres from our destination we ran out of fuel. Radio reception was bad and we could not get in touch with any of the camps. We decided that it was too late to start walking as it was almost dark. That night the temperature went down to −7°C. The only thing we had to protect us from the elements was a plastic tarpaulin under which we huddled together and a bottle of Old Brown Sherry. We tried burning a few bushes and grass but it didn't help for long. It was an endless night. We felt rather embarrassed for having treated our guest so badly, especially as we had also given a lift to a young chap who had all his belongings with him: he settled down on the back of the truck wrapped up in all his bedding and enjoyed a good sleep.

We had a few worrying moments in 1977 when the Parks Board considered the demand by the government to fence the Nossob River for security purposes, as this was the only international boundary in South Africa that was not fenced. It would have had a devastating effect on the wildlife. Despite the support for the fencing idea of some high-ranking officials in the Parks Board, the le Riche brothers were fortunately able to dissuade them. As a compromise, in early 1978 the powers that be decided instead to erect beacons along the Nossob River to demarcate the international boundary. Initially each beacon was to be surrounded by a security fence for some reason. We were unsure whether the final decision was made because of monetary restrictions or whether sense had prevailed, but in the end unfenced 60-centimetre-high markers were built at approximately one-kilometre intervals at the lowest point of the river-bed. In addition, by the end of 1977 Nossob and Twee Rivieren camps were surrounded by security fences, and by early 1980 a series of tall radio masts had been erected throughout the park as part of a security system. This theoretically meant that in the case of a terrorist attack helicopters could arrive within half an hour. Knowing how isolated the park was, we could not understand what all the fuss was about.

The year 1977 culminated in Gus completing his MSc on the food and feeding ecology of the brown hyena. I had typed the first drafts on my little manual typewriter, a twenty-first birthday present, but we decided for the final draft to go to Johannesburg and hire an electric machine, which was bliss to work on. The Parks Board gave Gus two weeks leave for this. He then had to go up to Pretoria in December for his oral, where he successfully defended his thesis. This was a relief, but the more daunting task of a PhD now lay ahead.

CHAPTER 3

Running
the Camps

As a scientist with National Parks Gus was entitled to have a technician working for him, I was keen to get the job, having worked as his unpaid technician for the previous three years. However, because of a number of logistic and administrative issues, complicated by the particular situation in the Kalahari, we were unfortunately unable to agree to the terms and conditions.

However, as the wife of a Parks Board employee living in an isolated location, I was obliged to help out with the running of the tourist camps when the camp managers (the ranger's wives) went on leave or had to attend to other matters. While I was happy to assist I found it a rather frustrating and thankless job. It meant that I was tied to the camp and could not go out at all with Gus. It was fine for a few days at a time, but when it stretched into months I was ready to leave the Kalahari. And of course, trying to run a tourist camp far away from anywhere involved many logistical problems and there was always some drama to cope with. Our only form of contact with the other camps was the radio, and often the reception was so bad that we could not get our messages across. I have an everlasting memory of having to say, over and over again, slowly and articulately, 'Kan ... jy ... my ... hoor? ... Oor.' (Can you hear me? Over.)

In October 1974 I was asked to run Mata Mata Camp for the princely sum of R150 for the month, having to be on duty 24/7. (Gus's monthly salary had just been increased to R417, which was still very little for a professional.) Tourists moving around the park had to notify each camp when they were leaving and where they were going for the day. Each evening camp staff would check that everyone had reached their destination. If this were not done, visitors who got stuck or had car trouble (or went on roads that they weren't allowed to) might wait for days before being rescued, such were the distances between camps and the small numbers of tourists. One evening three women didn't come back to camp. Luckily for me Gus was also staying at Mata Mata as he was studying the hyenas in the Auob River. It fell on him to set out and hunt for the tourists, while Stoffel le Riche set out from Twee Rivieren to search from that side. The three were eventually found, stuck on the dune road.

The roads between Twee Rivieren and Mata Mata were often used as a through route into South West Africa. This caused many problems. The park rule was that visitors had to take two and a half hours to get from the one camp to the other. The time of departure from a camp was written on the permit and was checked at the exit camp. One day a couple arrived after having only taken 80 minutes to do the 120-kilometre journey. They wanted to lynch me when I made them sit at the gate until the minimum time was up. Once Gus was driving in the Auob and a car came up very fast behind him. As he stuck at 50 kilometres an hour and would not let them pass, the tourists just zipped across the river onto the parallel road along the river and sped away. Luckily they were caught at Mata Mata and given a R10 fine by the ranger.

At certain times of the year groups of hunters would travel through the park and were, more often than not, very well oiled. They did not appreciate the two-and-a-half-hour minimum time limit. We learnt that they would drive like crazy to the picnic spot about half way up the Auob, spend an hour or so having a *skottelbraai*, and then drive enthusiastically to the exit gate, having thus passed the minimum time limit for the whole trip. Unfortunately as there were not enough staff members to monitor this practice, they were rarely caught. Their firearms had to be sealed on entering the park and unsealed at the exit gate. On several occasions at Mata Mata I felt very intimidated as a young girl on my own having to deal with foul-mouthed, drunk men brandishing firearms. One time I even had them rubbing my stomach when I was very pregnant with our first child and was helpless to do anything about it. I hated being there on my own and was thankful when in the 1980s Mata Mata was closed as an exit gate into South West Africa. It has since been reopened, but visitors now using that route have to spend at least two days in the park, so that problem has been solved.

It was much easier running Nossob Camp because when it was quiet I could nip home and get on with chores there. But we still had plenty of fun and games. Several times I had to deal with double bookings in the camp and ended up having strangers to stay in our house. A Mr Albutt and his wife, and their parson and his wife, arrived one day when the camp was full with a booking only for the following night. It was too late for them to go all the way back to Twee Rivieren. I took them home and quickly moved out of my room (Gus was away at the time) and made up two other beds. I chuckled at the thought of Gus returning home unexpectedly in the middle of the night and jumping into bed with strangers.

At about 8.30 pm a large, highly agitated German tourist came over to the house brandishing a spanner. Apparently there was a water leak in his cottage. Mr Albutt, the parson and I went over to the cottage to find water flowing out from under the doors. There were six people in the cottage and apparently they had all been in bed when this happened. The big guy had gone to work with his spanner on some of the pipes, but to no avail. Even though the main tap had been switched off, water was still pouring out of the walls.

When I think back, it was quite funny, but at that stage it wasn't a joke. We all stood around trying to puzzle out where the water was coming from. Then I remembered that some builders had arrived that afternoon. About six weeks previously they had put new pipes into the cottage. We called the man in charge, and when he saw all the pipes that the German had tampered with hanging from the walls, his face dropped. He went to another tap outside the huts next door and turned that off, and the water at last stopped flowing. Apparently the old pipes were still in the wall and someone must have turned the old 'mains' open, causing the pipes to burst. Anyway, just before 10 pm we got the place dried out and came home.

Once we had people camping in our garden and using our ablutions because the camping ground was too full for their liking. We had visitors at the time and found it rather embarrassing eating our meal inside and having strangers come in to use the bathroom. Today I wouldn't tolerate such demands from tourists. Another time, after an incredible storm in which we had 22 millimetres of rain in 20 minutes and hailstones the size of marbles, there was a knock on the door and I found three rather bedraggled campers who were friends of friends. They ended up spending the following two nights with us. Luckily I had enough dry bedding to provide for them.

I dreaded the times when camp machinery and equipment broke down, especially when visitors took out their frustrations on me. While I understood their feelings when things didn't work properly, for me sitting so far away from any help life became extremely difficult. At one stage I had been running Nossob Camp for several weeks and had had one crisis after another. For a whole week none of the geysers had worked and tourists quite rightly complained bitterly, especially as the temperature went down to freezing every night. After bursting into tears when 'attacked' by a particularly irate visitor, I got fed up and sent a message to Elias, who was now the warden based at Twee Rivieren, saying that if he didn't send someone immediately I would return the keys to him and have nothing more to do with the running of the camp. The response was that for five days I had three handymen working hard trying to fix the geysers and leaking taps and toilets. I couldn't even invite the tourists to come and shower in our house as our geyser wasn't working properly either. The drains had also been blocked and a machine had to be driven up from Twee Rivieren to solve the problem.

Apart from these camp maintenance issues, I suppose the thing I disliked most about running the camps was the poor communication with the authorities. Once a Mr Asham and his daughter flew themselves to Nossob, having being told by head office that there would be a vehicle waiting for them and someone to look after them. We had been told nothing of this at Nossob. Once we recovered from the surprise of their arrival, we discovered that the ranger from Mata Mata was supposed to act as their chauffeur and he had been given different dates. Of course it fell to Gus to drive them around. The Ashams had no food either so came and ate with us for two nights. They were very pleasant people and we met up with them several times after that, but once again there had been a serious lack of communication and a lack of appreciation from head office of the difficulties of our living in such a remote place.

On another occasion a Parks Board official flew in from head office with seven journalists on a Kalahari promotion exercise. As there was no official vehicle, our Kombi was commandeered to drive them around. We didn't mind, but sometimes people took things for granted. At least the 'VIP visitors' had started bringing most of their own food and we didn't have to provide everything for them. Twice Gus was told (never really asked) that he had to do relief work at Mata Mata. Parks Board apparently did not have the money to employ me instead. Though Gus was not charmed, he made the most of it by catching up on the analysis of his data and the writing of scientific papers in between checking in visitors.

In early 1980 the park warden Stoffel le Riche died of a heart attack at the age of 44. His brother Elias and wife Doempie then moved down to Twee Rivieren to take charge there. I was asked if I would take over the running of Nossob Camp on a permanent basis but had no hesitation in replying with an emphatic no. I did agree, though, to be in charge for several months until they appointed a new ranger and camp manager. Div and Alta de Villiers and their three offspring aged ten, eight and six arrived in June of that year. We spent a happy few years with them as our neighbours. Their children had to go to school in Askham, a little settlement 70 kilometres from the entrance to the park, and only came home once a month. Alta found it extremely difficult being separated from her offspring and I admired the way she coped. Our children and theirs always played happily together in the holidays.

I was very upset in the middle of 1983 when I suddenly found I had to run the camp and was not able to go on a long-planned family holiday to South West Africa, where we were to join Gus's folks. We had children by then and wanted to give them a seaside holiday. Stevie (nearly two) and I stayed behind while Gus and Michael (who was going on for four) set off for Swakopmund. It was an awfully long, lonely week for me and I was smouldering most of the time. One night a car hadn't returned to camp, so at 6 pm I quickly went home and bathed and fed Stevie and then set off in the Land Cruiser. Just out of the gate I decided I needed a tow rope and turned around to get one. Just then in the darkness I heard a 'yoo-hoo': the lost couple had walked about five kilometres back to camp. Two male tourists then kindly drove with me to the car. Its back axle had broken and the wheels were so splayed out that we couldn't tow it. The AA was called in to sort out the problem, but as you can imagine it took several days to do so.

A delightful young girl by the name of Elsie Pojani worked for us for a while and was a great help with the boys, especially when I had to run the camp. One morning she informed me that the following day she was going to marry Frikkie, who worked in the camp. The priest from the nearest settlement (about 200 kilometres away) was coming to perform the ceremony. She had nothing suitable to wear. Luckily I had a white dress that I could lengthen for her, plus some white stockings. I then baked a chocolate cake (not really my forte) and decorated it with white icing, flowers and their names. She was thrilled. The ceremony, to which we were invited, took place in an empty staff house and Frikkie looked dashing in a brightly checked suit. We also presented them with some sheets, pillowcases, towels and a tablecloth that we had been given as wedding presents and that were still in their wrappings. Elsie and I used to spend time together in the evenings making children's soft toys and other handicrafts when Gus was away. I always appreciated her company.

CHAPTER 4

Kids in the Kalahari

THOUGH WE WERE VERY HAPPY with our lives and lifestyle in the Kalahari, we suddenly realised that time was marching on and we had been married for nearly six years. It was time to make a decision. Did we or did we not want to have children? I have always loved children and my mother was surprised when I didn't fall pregnant in our first year of marriage; in fact I think she thought there was something wrong with us. But if we had babies there was no way that we could continue to do what we were doing, so we just pushed the question to the back of our minds. Yet the thought of getting old with no children and grandchildren around us was not at all appealing.

So in February 1978 a visit to the doctor in Johannesburg confirmed that I was pregnant. I was told that I should have monthly check-ups. All the same, the idea of driving 420 kilometres to the nearest doctor in Upington, over badly corrugated roads, did not make any sense to me, particularly as I was feeling well. So I next saw a doctor in Grahamstown at the end of April. Both doctors strongly recommended that I spend my last month of pregnancy near medical assistance, particularly as it was my first pregnancy. Gus did not feel that he could take a whole month off before the birth, so we decided that I should go to his parents in Johannesburg and he would come a few days before the supposed D-Day.

Getting to Johannesburg was a problem. I did not want to drive all that way by myself and South African Airways would not allow anyone to fly who was more than seven months pregnant. With four weeks to go, I managed to get a lift from Twee Rivieren in a six-seater Cessna with four male strangers. It was the worst flight of my life. I hadn't taken into account that at eight months a pregnant mother needs to urinate frequently. Even though I found a bush to hide behind just before takeoff, four hours of pressure on my bladder caused me intense pain and by the time we arrived at Lanseria airport I was in agony and in tears. A couple of days later I decided to do my six-monthly shopping stint before the baby arrived, and as my parents-in-law were unavailable, I set out on my own. The whole episode of packing and carrying heavily packed boxes brought me into labour three weeks early. Gus was blissfully ignorant of the goings on, following the brown hyenas deep in the Kalahari.

My parents-in-law managed to get a telephone message through to Izak Meyer, the ranger at Mata Mata, having had no luck with the phones at Twee Rivieren. He was adamant that it was a false alarm and so did not transmit the message to Nossob the next morning over the radio. Only when he was informed that Gus was now a father did he radio Nossob and the search for Gus started. Michael was three days old before Gus met him in the Marymount Hospital and it was another four days before he was allowed to hold our precious bundle, so barbaric were the rules in maternity homes in those days. Gus could only look at him through a glass window, which drove him crazy.

Although the process of having a baby living so far away from a hospital was challenging, it was nothing compared to what the local inhabitants had to cope with. Twice I had to act as midwife at Nossob because of premature births. In the first case the mother was not prepared for the event in any way. Luckily I had children by that time and so dashed home to get blankets, clothes and some scissors to cut the cord, as the ones they had were blunt. After a couple of hours she produced a healthy son. It was an emotional time for all of us and we were relieved that it all went well.

We were not so lucky the second time, even though we managed to locate a doctor in the camp. Not five minutes after we arrived, the baby appeared. It only weighed 200 grams (on my kitchen scale) but only lived for three hours. Its lungs weren't properly developed and would only take a gasp every 30 seconds or so. Luckily the placenta came away easily and the mother survived the ordeal.

Our lives certainly changed dramatically with the arrival of our first child. We got the first taste of it during the final months of my pregnancy when I found that sitting in a vehicle for long hours bouncing over the dunes was not making me feel too good. The only sensible thing to do was for me to stay at home and Gus to go off on his own. It became very lonely at times after that. In the meantime I was preparing myself for motherhood by reading as many books on baby care and child raising as I could lay my hands on, decorating a baby room and making all sorts of baby goodies.

A friend once introduced us as 'Gus and Margie – they live 100 miles north of F***-all'. There were times when it did feel like that, especially when the children fell sick, as I was soon to find out. Ten days after Michael was born, my mother arrived from Harare and then the three of us flew to Upington where Gus picked us up for the last leg of the journey back home. Our vehicle was a Toyota Hilux single cab, and the nine-hour journey back to Nossob with three adults in the front, a carry cot with baby on our laps, on dusty, corrugated roads, with the temperature over 38°C, made me declare that I would not be leaving Nossob again until we had a bigger vehicle.

No sooner had we settled down at home than Michael developed what we later found out was an anal fistula. When I first saw it, and my mother, who had had six children, did not know what it was, I spent an agonising night thinking terrible things. At first light I went around the tourist camp, knocking on doors in search of a doctor, and was mighty relieved to find one. He informed me that it was not life-threatening but that we would have to do something about it. Michael ended up having an operation at 10 weeks to sort out the problem. We returned home in a brand-new, out-of-the-box Volkswagen Kombi that cost R7 000, which was to be an absolute joy for many years.

Back on my own without my mother, I soon realised that it was best to follow a 'mother's intuition'. Luckily, Michael was on the whole pretty healthy. On the few occasions that he really needed a doctor, there was one in camp. Once we had a veterinarian look at Michael and give advice. I kept a fairly comprehensive medical kit, so had antibiotics and the basics at hand. But I cannot deny that I found it particularly stressful when the children were sick and agonised over whether we needed to travel vast distances to get help. I felt more comforted when Elias learnt to fly and there was a plane based at Nossob for a short while.

Because I was now stuck in camp most of the time with a little baby and spent a lot of time on my own, Gus bought me a double-bed knitting machine. Over the years I knitted literally hundreds of jerseys for anyone and everyone. I was in fact always busy with many varieties of handicrafts, something I have always loved to do. I was still very involved with the scientific project, keeping up to date with plotting the movements of the hyenas, the faecal analysis, game count analysis, typing and so on. I don't remember the time ever dragging or wondering what I should do with myself. I felt that the way not to feel lonely was always to be busy.

Summer-time for a baby in the Kalahari is very difficult and ours suffered terribly from itchy heat rashes. One solution was to lie them down on a wet towel and continually dampen it. I had constructed our own pram out of a cane baby crib, which I mounted onto wheels and then covered on the inside with a thin material, allowing for good ventilation. The conventional prams would have been far too hot. When the children were bigger and were running around, I used to peg wet flannels around their necks where the itchy rashes troubled them the most. Scorpions were a worry at night as they could climb walls and get through windows. I twice found one under the plug when I was about to give the baby a wash in the basin. For this reason I always used to cover the pram and cot with nets.

If organised beforehand, inoculations could be done at the clinic at Grenspos, which meant a journey of only 420 kilometres there and back. When I developed mastitis and felt achy and sick, I had to get Doempie to radio Judith at Twee Rivieren and ask her to phone a doctor in Upington to enquire whether I could take the antibiotics I had in stock while still breast-feeding. While he said I must take Michael off the breast, one of my baby books said he must continue drinking and I should use hot cabbage leaves on the affected area. Luckily, cabbages are particularly long-lasting and I still had part of one in the fridge. So I let him continue drinking and, with the help of the cabbage leaves, we both survived. I was not keen to stop: no need for sterilisation or heating, and the logistics of getting a constant supply of baby formula to Nossob was too daunting. Neither of our boys ever used bottles: they went straight on to baby cups.

It always amused me that, when our babies cried in the night, Gus professed never to hear them, and it was always my job to get up and sort them out. But when a spotted hyena called far in the distance, Gus would wake immediately and enthuse about the 'music in his ears'. He did rationalise by saying that, not possessing a milk bar, there was no point in his being disturbed from his slumbers. I had to agree he had a point. In all other ways he was an excellent father and did not shy away from changing dirty nappies and similar parental duties.

At the end of 1979 I thought I might be pregnant again, even though I was still breast-feeding. I had a pregnancy test after Christmas and it came back negative, so I forgot all about it. As the months passed I became worried about myself, knowing that something was not right; every now and then I could even feel a lump. Eventually we went to Upington and Dr De Klerk just laughed when I told him that something was wrong with me. I was four months pregnant and felt embarrassed that two biologists had not realised it. Gus had even said that I was beginning to look bloated. I had taken the result of the pregnancy test as gospel.

I was thrilled to have Gus at my side when giving birth to Stevie in the Settlers Hospital in Grahamstown. We went there because my parents had very sadly had to leave Zimbabwe and were now stationed there. I had only met the delivery doctor the previous afternoon. As with my pregnancy with Michael, I managed to have a check-up only twice before giving birth. Gus was even allowed to cut the cord, which was a far cry from our experience with Michael. It was wonderfully emotional. Follow-ups at clinics were impossible. As I could see that my babies were growing nicely, it didn't really worry me.

The children and I would join Gus in the bush whenever possible. When the hyenas had a kill in the river I would take the Kombi and watch from there with the boys, while Gus stood alongside us in his work vehicle. After one such evening I told our parents: 'I was so excited to be out again and we saw seven different brownies. A spottie came to the carcass but Normali stood her ground – a fantastic evening.' It was outings like this that kept me happy. The boys would sleep and play contentedly in the vehicle and also really enjoyed watching the hyenas or anything else we saw.

I don't think there are many children in the world whose first sounds they utter are hyena whoops. Michael used to love 'being a hyena', and whenever he wanted to call me he would whoop. A loud whoop signalled the end of a toilet session, which always amused our visitors. Stevie always used to try to emulate his big brother. It was wonderful watching their reaction to the hyenas. The boys loved being with them. Michael told them off on many occasions when they chased jackals away from kills: he didn't like bullies and wanted everything to be shared.

Sometimes we took the children to a brown hyena den, Michael sitting between Gus and me in his car seat and Stevie on my lap, our vehicle being a single cab. After an especially exciting excursion I wrote: 'Chinki arrived and called two little ones out of the den. Mickie kept doing a spottie call but it didn't seem to worry them. On the way back we came across a mother genet and her two large offspring foraging in a tree. It was amazing to witness them gliding through all those thorns with no difficulty whatsoever. They looked very vulnerable running across the open ground when moving from tree to tree. We then came across a honey badger trotting along out in the open, and just before we got back to Nossob we found three adult wild dogs with two pups feeding off a springbok. I need this kind of outing to keep me sane.'

Once the boys began to play outside I became concerned about the cobras and puff adders that we frequently saw in the garden. I did not want to take a chance, being so isolated. When there was no one in camp capable of catching and removing them, I had them killed, much to Gus's disgust. I always got into trouble with him for doing so and I did feel bad about it, but I would never have forgiven myself if either of our children had been bitten.

My main concern about bringing up our boys in such isolation was that they would not learn to interact socially with other children and with people in general. While they were growing up I did my utmost to let them have as much contact with others as possible. Some of the camp staff had little ones and they would come and play on a regular basis. Michael was never good at entertaining himself and always needed someone to be with him. I spent countless hours being his playmate, building cities out of mud and playing cars in the sand in the backyard. By two and a half years of age he would dash out and greet any arriving tourists like long-lost friends and often came home with oranges and other goodies that he had been given. It was one way of getting fresh food. He got particularly excited when there were children in the vehicles, and quickly called me so that I could suggest to the parents that I look after their offspring if they got tired of sitting in the car viewing game. This offer was often taken up.

I often worried about our children not having enough fresh fruit and vegetables to eat. Any visitors knew that they would be welcomed with open arms if they arrived heavily laden with fresh things. When the boys were babies and we went to town, I would bring home vast quantities of fresh beans, pumpkin, potatoes and the like, cook and purée them and then freeze them in meal-size portions. I can remember becoming very possessive about our fresh food and getting upset (although hopefully not showing it) when our house visitors would (to my mind) thoughtlessly eat two precious apples in one day. I always rationed ourselves to make the fresh food last as long as possible so that we would not have to go too long without.

When Stevie was about a year old he suddenly developed horrid-looking yellowish, crusty sores on his face, then on his fingers, elbows and knees. At that stage Michael had for company three little girls belonging to the camp workers. I soon discovered that the eldest of the three had similar sores covering her bottom. Luckily I found a doctor in camp and he diagnosed the problem as impetigo, a skin infection which is caused by bacteria that lives in the nose and is very contagious. I now had the huge job of sterilising the whole house and all its contents, plus the Kombi, as I had taken all the kids for drives. The doctor put little Anna onto antibiotics as she had the infection really badly and, if left untreated, could have resulted in kidney failure. We all had to put special cream up our noses and wash with medicated soap and be careful not to touch the sores. I got one on my elbow and found it pretty painful. Just when I thought we had got rid of the outbreak, I discovered that little Sissy, one of the camp workers' daughters, had developed sores on her feet. As she had been wandering through the whole house, I had to start all over again with the sterilisation process. It took well over a month to finally get rid of the infection.

Michael once swallowed a coin and we were relieved the next day, after careful examination, to find it. Gus wrote to his parents, 'What a good job Margie has had such vast experience in these matters from examining the hundreds of hyena scats!' I know I would have worried about blockages if we had not found it and would have felt that we must make the long trip into Upington to consult a doctor.

Reading through all my 12 years of letters to our parents, I found only a few times when I complained about living where we did. An example: 'I must admit I dislike Sundays when I'm on my own, in fact I think I need a break from this place. I seem to be in the house for 23½ hours a day, which isn't exactly good for me. The children get fed up in the car after 10 minutes and then it isn't a pleasure to go for a drive anymore. Besides it has been pretty hot and dusty and there hasn't been much around to go and look at. No doubt I will survive, but it would be nice to have a few family and friends around today and to do something really different.' Fortunately the Kalahari wasn't as bad for me as for one tourist I dealt with: she threatened her husband with divorce if he did not immediately take her back home after the second day of their week-long visit.

When Michael turned three, I signed up with the '345 Nursery Group' programme from England. Every month I was sent books and activities to do with the children and religiously spent two hours a day working through the course with the boys and any local children that were around. They really loved it and were always excited to learn, and I felt it was good to have a bit more structure to the boys' day. It also made my life easier to have this source of ideas to assist me in keeping the children busy and happy. Living so far away from shops, I quickly became a hoarder, always imagining some use for an old box or an empty toilet roll. We spent endless hours creating cars, houses and the like out of old boxes. In fact the only 'toys' we bought for our children were sets of Lego; the rest we made ourselves.

When Gus was working on the spotted hyenas, he usually went away for the whole week as it was too far for him to come back to camp each night. It would have been too time-consuming and costly. His bush home was a three-sided structure of wire netting that had been made for the lion-capturing programme. Though it did have a door that enclosed the fourth side, Gus didn't think it was a necessary to use it. He positioned it in the shade near where the main clan was living, about 50 kilometres from Nossob Camp. I would often drive up with the boys in the Kombi in the middle of the week, arriving at the 'bush camp' at about 2 pm, by which time Gus had hopefully had enough sleep. We would then spend an hour or two together, have a braai, and then I would return with the boys while Gus and his assistant Hermanus would continue with the research. We really looked forward to these outings and it helped to make the lonely week go past.

Once I had an incredible journey there in a thunderstorm. The road was a flowing river in places and I had to keep 'bundu bashing' in the poor Kombi. On that occasion Gus actually came looking for us as we did not arrive at the usual time. I was mighty pleased to see him. I never went out of Nossob Camp without taking a good supply of food and drinks in case we got stuck for any length of time.

One night at home Gus had a nightmare and let out the most ghastly blood-curdling scream I'd ever heard. I thought he was being attacked by someone. He apparently dreamt that a spotted hyena had him by the shoulder. I made him promise me there and then that from that moment on, when he slept in his bush shelter, he would always securely fasten the door. I like to think he kept his promise, but I have my doubts.

Later on, when Gus was spending time with the spotted hyenas in the lower Nossob, he managed to borrow a caravan which he used as his daytime shelter. I took the boys there several times to spend the night sleeping in the caravan while Gus was out following hyenas. In the morning when Gus returned we would get the news of the night's activities; he would then jump into bed and we would return home. It was a treat for us all.

Once we became parents, one of the few times that I could accompany Gus on his nightly jaunts with the spotted hyenas was when our parents were visiting and they would stay with the boys. I treasured these occasions. One such time we all went up to the bush camp for a braai. My parents returned to Nossob with the boys while Gus and I went to the den to find seven adults and seven youngsters adorning the site. 'Just as the sun was setting all but the two babies set off from the den. It was a truly beautiful sight with the hyenas at one stage being silhouetted. Suddenly they started running at an incredible speed and we lost all but the two youngest ones, who after stopping and listening for a few seconds led us to the rest of the clan. They had surrounded a gemsbok that was standing in the middle of a treacherously thorny *Acacia hebeclada* clump about 10 metres in diameter. The hyenas did not linger, obviously realising that this animal was not going to fall prey to them. I found the chase extremely hair-raising but very exciting.

'Soon they started chasing after something again and we tore after them in the usual fashion. This time they had surrounded four female eland that were in a tight bunch, heads together, and when a hyena tried to go for them they'd kick like lightning with their back feet. From circling these eland they dashed off on another chase for one kilometre or so and we found them surrounding another female eland. Again the same thing happened, the hyenas trying to get a bite in and the eland kicking. By this time I had bitten off all the nails on one hand! The main thing that worried me on these chases was that we'd run over one of the little ones because we were literally moving as one of the pack.

'After circling this eland for a while, the hyenas then continued on their way for another couple of kilometres. Suddenly they all stopped and one of them started whooping. We then realised that one of the young hyenas was missing. After about 10 minutes of waiting it appeared making an incredible noise, obviously very pleased to be reunited with the clan. On they moved and it wasn't long before they gave chase after a gemsbok, which managed to back up against a bush and defend itself. The hyenas didn't spend long with it, obviously knowing that they had no chance of catching it.

'By this time, having travelled 17 kilometres, the biggest cub and one of the females and her two cubs had dropped out of the pack, so there were just eight animals present. At 27 kilometres the eight gave chase again and we most frustratingly lost them. We stopped on a high dune and played the tapes, but to no avail. We then went back to where we first lost them and started following their spoor. Gus sat on the front of the Cruiser with the spotlight, and to our utter amazement and delight after one kilometre we saw a whole bunch of eyes. The six adult hyenas were feeding on an adult female gemsbok. Gus was so upset that we'd missed the kill as we had never seen them pull down an adult gemsbok. Anyway, we were just thankful that we had found them again.

'I was wondering what had happened to the two young hyenas that had been with the adults at the beginning of the chase. After a couple of hours the female and her two cubs that had dropped behind, arrived on the scene and then just before we left at 3.30 am the last two appeared. Their mother and one of the males had whooped for them a few times. Whether they got lost or just lagged behind, we don't know. I found it quite incredible that the young ones managed to keep up with the adults all that time and distance. And to think that they had to travel about 27 kilometres to get back to their den. It is certainly a night I'll never forget and I must admit that this one night with the spotties was far more exciting than any one night I had with the brownies. We took two hours to get back to camp by which time Stevie needed a feed.'

CHAPTER 5

Wild Visitors
to the Camp

WE WERE VISITED DAILY by several yellow mongooses and had a resident colony of ground squirrels that entertained us. Both species very quickly becoming tame. They were the closest things we had to pets. Only the rangers were allowed to keep dogs, ostensibly for work purposes. Gus was adamant that we would never take in 'orphaned' wild animals: it was lovely when they were small but the association only ended in tears once they became adults and had to be got rid of. Although we were surrounded by animals I felt that our children lost out by not being able to have any physical contact with them, as one does with a domestic pet.

On my parents' first visit after we had moved into the house, my very creative mother built us a large, beautiful-looking bird bath in the front garden using cement and softening the lines with the local, interestingly-shaped calcrete rocks. Over the years it provided many wonderful sightings. Although we recorded only 90 species of birds in our garden, many appeared in vast flocks, especially in the winter. Little banded goshawks caused havoc with flocks of grey-headed sparrows, shaft-tailed whydahs, scaly-feathered finches, masked weavers and red-headed finches. Two pearl-spotted owls were also regular visitors.

Gus has a theory that if one stays in one place for long enough in southern Africa, one is likely to see every bird in *Roberts*, because birds can fly. We certainly had some strange ones blown in. A palm-nut vulture stationed itself at the bird bath for several days, we found a buff-spotted flufftail in the backyard and an African crake took up residence in the garden for a week. Ruffs visited a couple of times and a pink-backed pelican was seen on the landing strip. On one visit my mother told Gus that she had heard a paradise-flycatcher in the garden. Being the sceptic that he is, Gus didn't believe her. She took great pleasure in finding him the bird a few hours later. A strange sight was that of a few night herons roosting in one of the garden trees. They came back for a few days and then disappeared. I found a dabchick near the office, caught it, put it in a box, then read up in *Roberts* that it feeds on frogs and water insects, so realised that we had a problem on our hands. How it got here in the first place we just didn't know – they breed in open stretches of water and there wasn't much of that around. We then put it in the bird bath, hoping that it would find something there to eat. After paddling around for a while it got out and then, much to our amazement, after a fairly long run, managed to get up into the air and flew off to the river, out of sight. I must admit I was rather relieved to see it go and just hoped that it found its way back to where it came from. The thought of having to search for frogs in this environment was rather daunting, to put it mildly.

Our first really unusual bird sighting took place while we were still in the caravan. We found a dead seabird under it. Not having the facilities to freeze the entire bird, we stored the legs with their partially webbed feet in the tiny freezer compartment of our fridge. A short while later, our friend the bird expert Richard Liversidge was visiting and, after examining the feet and from our description of the bird, he was pretty sure it was a phalarope. We kept the legs because they were unusual and some time later, when we wanted to show them to somebody else, we found they were gone and wondered whose meal they had unwittingly embellished.

Other unusual ornithological treats were a striped kingfisher, a painted snipe, a juvenile saddle-billed stork, a gymnogene and a single-helmeted guineafowl which we saw several times over a period of a few months at Kaspersdraai Waterhole. Perhaps the most incongruous sighting was six white-faced ducks walking down the road towards Cubitje Quap.

Once while following a brown hyena in the river-bed, we stopped under a small camelthorn tree to see if we could find what the hyena had been investigating. We couldn't discover anything but looked up to find 17 swallow-tailed bee-eaters perched in two rows of eight and nine. That beautiful sight made an indelible imprint on my mind.

For a year we had the fun of a tame secretarybird roaming camp. It was picked up by the ranger when very young and hand-reared. Watching Seckie dispose of his favourite food, snakes, enthralled us and visitors alike. His lightning-fast reactions and agility were astounding to watch. His antics when learning to fly often had us in hysterics. He like to roost on top of one of the huts at night and, as dusk approached, he would nonchalantly make his way to the top of a slope in the far corner of the camp, and then run down with all the speed he could muster to get airborne. His landings on the roof were very amateurish, and often he couldn't keep his balance and half fell, half flew off to try again. As with most 'tame' wild animals, the bigger he got the more problems he caused, and he started chasing children in the camp. I was luckily close at hand when he jumped up and tried to claw at my one-year-old niece's face. I managed to grab his tail and pull him away, preventing him from causing any damage with his razor-sharp talons. Shortly afterwards he disappeared, hopefully enticed away by a mate to live happily ever after.

There was a rather eccentric old lady visiting Nossob Camp who was keen to take photos of the gorgeous crimson-breasted shrikes which are notorious for never staying still for a second. With much palaver, she erected a hide and deposited some minced meat in front of it to entice the birds. After a few hours during which nothing happened, who should come along but dear old Seckie? Much to the photographer's disgust, he promptly ate the bait, and so she had to be content with photos of a secretarybird instead.

At the back of our house we had a large thatched rondavel that was surrounded by a low wall and chicken wire-netting to the roof with a gap in it for an entrance. In summer we decided to sleep in the rondavel as the house remained unbearably hot in the evening, its bricks not easily releasing the day's heat. On the second night we had blown out the paraffin lamp and were settling down in our new, low-slung beds, Gus nearly asleep, when I heard something rustling close to my bed. I grabbed the torch from under my pillow and shone it around, under boxes and other items lying around, but couldn't see anything. I was told by my better half that it was probably only a gecko, and that I should switch off the torch and go to sleep. I tried to obey, but the noise persisted. Flicking on the torch a second time, I was horrified to find a yellow cobra on the floor at the head of my bed not a metre away from me. The noise that I had been hearing was of the snake crawling through the large plastic bag that had covered the new mattress. I was out of there like a shot and, from the safety of the dining-room window, yelled to Gus that he was sleeping with a cobra. He nonchalantly shooed it out (luckily they are not aggressive snakes) and went back to sleep. I informed him that I was going to finish the night inside, and in fact this was the last time I slept outside at Nossob. I would rather deal with the heat problem inside than brave the creepy crawlies outside, something Gus could never understand.

Another reason why I was scared to sleep outside was that we regularly had a leopard visiting the camp, probably lured by the ranger's chickens and dogs. One morning Elias and Doempie woke up to find their precious little pet dog hanging in a nearby tree. On several occasions a leopard managed to get into their chicken cages and caused havoc. One night a female leopard and her cub also came to drink at our bird bath. We found their spoor the next morning right underneath our bedroom window, which had been wide open. The next night I surreptitiously closed the windows a bit when Gus wasn't looking. Lying in bed on another night, I heard lapping at the bird bath. The light of a torch revealed another leopard quenching its thirst. I promptly closed the bedroom door as all the house doors were open – but so were all the windows. It was really a pointless move.

One of the most exciting incidents in camp occurred while I was in charge. It was the winter and I had to go in the dark to the office for the 7 am radio session. As I was watching the dawn break I was startled to see, silhouetted in the large camelthorn tree in front of the office, a springbok carcass. There was only one way it could have got there. Gus had shot the springbok two days previously for baiting his hyena traps and had left it in the garden overnight. The springbok was duly removed from the tree and Gus went out to check his traps, taking it with him in case any had to be re-baited. In the middle of the afternoon he returned, having caught and marked a brown hyena. He asked his assistant to put the remains of the springbok in a little tin storeroom in our backyard to keep it safe from the leopard, who would likely return that night. The assistant was busy packing equipment away in the shed when suddenly from behind a box a baby leopard of about six months sprang out. I was on the shop veranda talking to some tourists and saw this spotted body come bounding through our back gate towards us. I grabbed Doempie's dog and threw it and the tourists into the office and slammed the door, for my first impression was that it was an adult. I then saw two chaps chasing after it and knew that no one in his right mind would be running after an adult leopard.

They managed to guide it back into our yard and cornered it against the house. It was all teeth and claws, hissing ferociously. Gus threw his hyena weighing net over it, and managed to pick it up by the scruff of its neck. It now hung like a rag doll, calm and quiet. This is how it was carried by its mother so the grasp was obviously comforting. As they let it go in the river-bed, it actually charged briefly before running off – it sure knew it was a leopard. While all this was taking place, I had to stand at the gate to prevent the now very inquisitive tourists from coming in and getting too close. We then checked the camp for the mother's spoor and found where she had left the camp. The two of them must have been disturbed when I went early to the office. While the mother could find her way out, the cub obviously got separated from her and hid away in the storeroom. All this time friends of ours, Shirley and Emile, were quietly taking photos of the whole episode, so we have a lovely reminder of it all. The following evening we saw that the leopards had been reunited.

At about this time we started regularly to see spoor of our favourite animal around the house. In one of my letters I wrote: 'It looks like some trickster is at work on our front lawn tonight. There is a large piece of hide with a long string attached to it leading to the bedroom window, at which stands a camera on a tripod. We are determined to see which of our friends is coming to visit us each night. We haven't quite worked out who is going to be tied to the other end of the cord (or what part of the anatomy) – should be fun to see my better half disappearing out of the window – or maybe it will be me!' As Gus always slept on the window side of the bed, it made sense that he was the one to be connected and he duly wound the rope around his arm, not tying knots in case he didn't want to go where the cord was taking him. At the first tug in the middle of the night he sprang out of bed, crashed into the tripod, and made such a loud noise that the 'tugger' quickly disappeared. Two nights later there was another 'bite'. This time Gus got a striking photograph of a brownie looking towards the bedroom window, hide in its mouth with the shimmering cord spiralling towards the camera. It was Jo-Ro from the neighbouring Rooikop Clan.

Another time we found a brown hyena in camp. Gus decided that we must not look a gift horse in the mouth and promptly darted it. As it luckily ran back into our garden, we could work on it in privacy, marking it and taking all the necessary measurements before releasing it from the camp.

Once we had great excitement with three lions in camp. They came through a back gate that had inadvertently been left open, into our garden, having a drink at our bird bath on the way, and then sauntered up to the ranger's house where they attacked his Alsatian. They then moved down into the camp ground. One of the tourists in camp was a keen recorder of animal sounds. When she heard them roaring she thought they were on the other side of the fence and dashed out with all her equipment, only to have a lion jump out at her. Fortunately no one was hurt and she had a great story to tell.

CHAPTER 6

Unusual and Exciting Animal Sightings

APART FROM THE WONDERFUL HYENA experiences during our 12 years in the Kalahari we also had numerous unusual and exciting animal encounters. Here something unexpected is liable to emerge when you least expect it.

Wild dogs are extremely rare in the southern Kalahari and we saw them only on about 10 occasions. The most exciting of these was when following two spotted hyenas along the upper Nossob. At 4 am they suddenly darted off to the left. In the moonlight we saw 11 wild dogs moving towards them, followed by 13 three-month-old pups. The dogs quickly lost interest in the hyenas but were fascinated by the truck and came right up to it. We wondered if this was the first time they had ever encountered a vehicle. We stayed with them until the sun rose, deciding that on this rare occasion hyenas were not the most important animals in the world. At daybreak they moved back to the Botswana side where they flushed out, killed and devoured a steenbok, the adults standing back to allow the pups to feed first, an admirable trait in these fascinating creatures' behaviour. This group was never seen again in the park.

Once a lone female impala appeared in the Auob near Mata Mata. We have no idea where it came from but speculate that it escaped from a neighbouring farm where it had been brought in. Over the next four years it made its way slowly down the Auob and up the Nossob as far as Kwang Pan, always in the company of a herd of springbok. Being a head taller than its companions, it stuck out like a sore thumb. The last time we saw it was when we investigated a report of three wild dogs at Cubitje Quap. Amazingly, when we found them they were eating the impala.

Although baboons are resident in the far eastern Botswana side of the park, the South African side is too dry to support a resident population. However, we occasionally saw some stragglers. One night Gus and Hermanus were following spotted hyenas when the hyenas stopped at a tree and kept looking up into it. Thinking there must be a leopard with a kill, Gus was surprised to find eight baboons sleeping there. Even more exciting was our sighting of a troop of 33 on Kwang Pan. They were extremely wild and would not let us approach, probably never having seen a vehicle before. On several other occasions we saw single males, probably young animals looking for a new troop, along the Nossob river-bed.

Warthogs are another marginal species occasionally seen in the upper Nossob near Union's End. In high-rainfall years we have even seen them with young.

The rarest mammal sighting was made by the photographer and filmmaker Richard Goss. He told Gus that he had seen three banded mongooses just north of Nossob Camp. Jokingly, Gus said he did not believe him: this species had never been recorded in the park, the nearest resident population being several hundred kilometers to the north. Fortunately a week or two later he again saw the mongooses, this time going into a hole just before dark. We were at the hole at sunrise and, sure enough, three banded mongooses emerged. It is amazing that such small animals can move so far.

We had a wonderful sighting of a leopard at Kaspersdraai windmill chasing 11 bat-eared foxes. The latter then turned table on the leopard, running circles round it, holding their tails up in an inverted U and barking their high-pitched screech in protest. The leopard was so intimidated that it escaped its tormentors as quickly as it could by bounding into a tree. An amusing sequel to this observation was that a well-known ornithologist also happened to witness the event; back in camp he argued with us that it was a cheetah and not a leopard that was involved.

Another cat and dog incident, this time spiced with a hyena, occurred when we came across three jackals that had just caught a springhare. Suddenly out of the darkness a caracal charged in and scattered the jackals. They soon regrouped and tried to reclaim what was rightfully theirs. The cat stood its ground, hissing, spitting and clawing, and after a brief scrap the jackals finally retreated. Half an hour later poetic justice was administered when a brown hyena, with little resistance from the caracal, poached the remains.

The Kalahari is famous for its lions. Somehow their stature seems to be enhanced against the backdrop of this desert landscape. The sight of a big black-maned lion adorning a red dune is unforgettable, even for a hyenaophile. One evening when our attempt to locate one of our brown hyenas was proving unsuccessful, we came across the two well-known males of the Kwang pride, coded BC and BE by Professor Fritz Eloff from Pretoria University, who was studying Kalahari lions at the time. Even though, to our way of thinking, two lions in the hand are not worth one hyena in the bush, we decided for once to digress. There were many springbok with lambs in the area and, within minutes of our finding the lions, BE snapped up a lamb. One would think that the so-called king of beasts would not stoop to squabble over such a morsel, but immediately BC was after him. BE ran off but soon put the lamb down and tried to eat it. In a flash BC was on top of him, trying to get hold of the lamb, and grappling at BE's face and head with his huge paws as he did so. To us it appeared that the lions were bent on killing each other for such little gain and that they were definitely in the process of maiming each other. For several minutes they stayed locked in combat, growling and snarling. Suddenly BE managed to break away and quickly wolfed down the lamb. I was struck by the ferocity of the two animals and would have expected there to have been some animosity between them afterwards. Yet, like prize fighters who shake hands after a bout, within minutes the two were rubbing heads and then lay down together unscathed.

Early on in our stay we came across two cheetahs near Rooiputs waterhole in the lower Nossob. They were lying under a small camelthorn tree in the river-bed. As we approached, one of them, a large male, got up and moved off a short distance with a pronounced limp. We then noticed that the second animal, also a male, was badly injured. Its mouth was open and infested with flies, which it made no effort to remove, it had a dazed look in its eyes and made no attempt to move away even when we approached to within two metres of it. As it lay looking at us, it occasionally rolled over from one side to the other.

After watching them for some time we continued on our way and about one kilometre further we saw a large leopard. When we arrived back at Nossob Camp we reported the sighting to Elias, who then informed Stoffel at Twee Rivieren. The following day Stoffel went out with trackers and managed to reconstruct what had happened. Some time after we had left them, the cheetahs had moved down to the waterhole to drink. While they were drinking they were attacked by a leopard, in all probability the one we had seen earlier. The leopard killed the badly injured cheetah and carried it off, dragging it into a tree some 400 metres away. Later the second cheetah came to this tree but was chased away by the leopard.

Two days later we inspected the tree and found the cheetah still hanging there. Very little of the carcass had been eaten: only the soft parts on the underside and a few internal organs. A month later the remains were still hanging there and nothing further had been eaten. This was an unusual case of predation, but it is by no means unique. Carnivores quickly recognise when another carnivorous animal is sick or handicapped and will readily attack in these circumstances. In the animal world it is always advantageous to get rid of your competitors.

One weekend some tourists reported that they had seen two hartebeest with their horns locked. Gus and Elias went out and found them close to camp, the one animal dead with a broken neck and the other exhausted from trying to free itself. Both were young adult males. As this occurred during an ungulate collaring campaign in the park, Gus and Elias decided to make the most of the opportunity and, with the help of several field rangers, they managed to fit a collar to the weakened survivor, just by holding it down. They then easily disentangled the horns, and the lucky one ran off, falling over several times but slowly gaining its strength. What a freak accident! During the operation Gus quickly snapped a few photos of the incident, including a dramatic one of the live hartebeest trying to escape as they approached. He later entered this image, 'Locked Horns', in the 1982 Agfa Awards Photography Competition and came first. His prize was a trip to a national park in Canada, but as he was being sent to that country by the National Parks Board to attend a conference the following year, he asked for a payout instead, saying that he would definitely use the money for travel.

Shortly after this Richard Goss arrived at Nossob to assist in making a Kalahari film. He and Gus, both being keen photographers, often joked when out shooting that they had just taken the next winner of the Agfa Awards. As luck would have it, Richard won the competition the following year with a photograph of a magnificent Kalahari black-maned lion dragging a gemsbok down the river-bed in a dust storm. Gus won third prize in the same competition with a photo of spotted hyenas on a kill called 'A Bloody Mess'.

Richard's prize was a trip for two to Tiger Tops Reserve in Nepal, which he and Karen enjoyed as part of their honeymoon. Gus and I chaperoned them for a whole month, using Gus's prize money from the previous year. For a week we visited various tiger reserves in India, then moved on to Nepal for two weeks. We hiked up towards 'the Fish Tail' in the Himalayas and then rafted down the Trisuli River into Tiger Tops. Our last week was spent visiting reserves and historical sites in Sri Lanka. It was a truly memorable and exciting holiday. It was the first time I had ever left the boys, now five and three years old, and was not looking forward to being parted for so long. We dropped them off with Gus's parents in Johannesburg, who then took them down by train to my parents in Umtata, so they were well looked after and thoroughly spoilt by both sets of grandparents. At one stage Stevie asked if his 'Kalahari Mummy' was coming back. I must admit the first three weeks were fine because everything was new and exciting, but thereafter I was quite ready to come back and be a mummy again.

CHAPTER 7

Nature, Science and Emotions

On the surface it may seem that in our relationship Gus is the pragmatic scientist and I am the emotional anthropomorphist. Although this is to an extent true, the driving force behind what Gus has tried to achieve is a passion and feeling for nature. He loves nothing better than to watch animals in their natural environment and to learn about how they function. However, we believe that in order to understand nature it has to be put into a scientific framework. We have both learnt that as beautiful as nature is, it is also, to our anthropomorphic and Eurocentric way of thinking, cruel.

I have always struggled with the cruelty of nature and dreaded seeing kills, never being brave about it. The first kill I ever witnessed involved two black-backed jackals killing a springhare. As I told our parents, 'I had to close my eyes and block my ears as their death throes cry is haunting and soul piercing. Oh why can't all animals just eat grass?'

Witnessing my first spotted hyena kill wasn't so bad as I had anticipated, I think because of the build-up beforehand. We spent several hours at the den with the cubs and adults, with lots of interactions between them. The adults then performed their ritual greeting, which serves as a uniting factor, and set off on the night's jaunt. The fact that we had been with the little ones helped me – I kept thinking that the mother had to feed in order for them to survive. Luckily I never witnessed any lingeringly long hyena kills and the prey usually died quickly.

Once on the way up to Gus's camp the boys and I came across three very bloody and full cheetahs nibbling on a springbok lamb. After watching this for at least five minutes the lamb started bleating. I felt sick and made a rapid getaway.

As hard as I tried, I could not stop myself from becoming emotionally involved with the animals that we studied, and used to get panicky when we watched them in dangerous situations. Gus always tried to calm me down and tell me that we were there to watch what was happening and were not allowed to interfere in any way. I knew all this but couldn't help myself. All my years spent in the bush witnessing such incidents have certainly not hardened me.

One evening while following Normali, she came across a lioness with three small cubs on a kill. Normally in such circumstances a brown hyena would have a brief look at the situation and then move on, returning to scavenge the remains sometime later when the lions had left. On this occasion, for reasons best known to her, Normali lingered longer and even slowly approached closer to the tawny cat. I started to feel uneasy when she persisted in what I saw as dangerous behaviour and began to admonish her for her recklessness. She, after all, had three small cubs in the den. My uneasiness turned to near hysteria when with little warning the lioness charged her. Normali seemed a bit slow off the mark and the lioness started gaining on her. We followed them through our binoculars in the moonlight as Normali ran for her life down the river-bed. Just as they reached the limit at which our night vision could function, the lioness appeared to catch up. We heard Normali give a loud short growl and then there was silence. I was convinced that our favourite brown hyena had been assassinated. We waited a few minutes, then saw the lioness return to her carcass. Thank goodness, from her radio signal we could tell that Normali was moving rapidly away, so she had escaped. We were relieved when we caught up with her to find an uninjured, but hopefully wiser, hyena. Gus recorded the incident on his tape recorder as it was happening. When we listened to the tape the next day my frantic cries of 'Oh no, oh no, she's got her' could be heard.

Not long after Gus started working on the spotted hyenas, some tourists reported that they had seen one at Kwang with a gin trap on its foot and that it was in a very bad way. Gus quickly went up with Div and found the poor thing with its foot hanging on by a tendon or two. It was Simon, one of the sub-adult males from the Kousaunt Clan. Where he could have come across the trap was a mystery, as the South West African fence, the one place where we knew these traps occurred, was 50 kilometres away. It is difficult to believe that he could have travelled this distance with such a grotesque burden. Perhaps it had been lying somewhere around Kwang for years, left over from the days when hunting was common, before the park was proclaimed in 1931. Gus reluctantly decided to put him out of his misery. Although these animals are resilient and instances of hyenas surviving on three legs have been recorded, it seemed that a young male that would have to leave his clan one day and survive for a time by himself might not be able to do so with such a handicap.

Although poaching was not really a problem in the park there were a few incidents. A long-standing feud between a South West African commercial farming family and the rangers ended in a wild vehicle chase and shoot-out, which tragically resulted in the death of the farmer's 23-year-old son. Less hostile but still serious were incidents of subsistence farmers from Botswana poaching meat and skins. When caught, they were often given the option of working in the park for a few weeks or being handed over to the Botswana police. Once one of these poachers escaped and was recaught 25 kilometres away a few hours later, much the worse for wear, as he had taken nothing with him and it was extremely hot and dry. I felt sorry for him until I heard that he and his accomplices had trapped 25 jackals and killed 12 eland for their skins.

Working with wild animals one soon realises that not all of them have read 'the book'. One day we visited an old spotted hyena den and, while we were searching for any fresh signs of hyenas, I heard a cub sounding noise emanating from the hole. 'Knowing' that adults do not go down into their dens, I got down on all fours at the entrance of the hole, hoping to catch a glimpse of the little one. The next moment I found myself staring into a huge face. Luckily the hole was quite small and the adult had to belly-crawl out on its knees, giving me enough time to run away towards the car, although I felt I was going to lose large chunks of my *gluteus maximus* while doing so. The poor creature obviously had no intention of feasting on me because it ran off in the opposite direction for a few metres and then stopped and stared at me. I certainly felt a huge adrenaline surge.

That same day I had a second fright. We had stopped to examine a fire-damaged tree when two Namaqua doves landed on the ground not a metre away from me. I stood still and watched them displaying to each other, excited at being so close. A minute or so later I heard a loud swooshing noise and, in a flash, feathers were flying. A lanner falcon had dive-bombed the doves, only opening its wings at the last split second to stop it from ploughing into the ground and giving it control to catch its prey. At such close quarters I thought I was about to be attacked.

This was not nearly so dangerous as Gus experienced one night when he slept in Auob with Hermanus. As it was cold they decided to sleep on their stretchers near the fire instead of on the back of the truck. At 1 am Gus was woken by the sounds of a lion roaring in the far distance, but he dozed off again. About half an hour later he awoke to find a lion standing about three metres from the foot of his bed, looking at him. He said it was the first time he had ever looked *up* at a lion and he felt extremely small. He clapped his hands wildly and, in not such polite terms, asked the lion to leave. Luckily, after a split second, when Gus was not sure what was going to happen, it moved off about five metres and started to roar. It felt like the whole world shook with the power of that magnificent sound, but at least Gus then knew that he was not going to be attacked. The lion's roar was the first Gus's companion knew of the encounter. He proceeded to put some more wood on the fire and climbed back into his sleeping bag without saying a word to Gus. The lion continued on his way, still roaring. It had been a spine-chilling experience, which they laughed about the following morning. I know I would not have closed my eyes again that night, but the men had a good sleep after the interruption.

I was never happy when we had to capture the hyenas and was always concerned that we were doing them harm. I knew that we could not conduct the project without marking and collaring some animals, so had to accept it. In the beginning, when some of the methods we used were new and untried, we had a few awful incidents but we quickly learnt what the best method was. It has always been uppermost in our minds that we remain as unintrusive in the animals' lives as possible.

On the work-front Gus was involved with the biannual aerial censuses, and he conducted monthly game counts along the river-beds. One amazing sight in 1979 was of an estimated 100 000 wildebeest on the South African side of the park. The capturing and collaring of wildebeest, gemsbok, hartebeest, eland and springbok and their subsequent monitoring took up a lot of Gus's time. The rest was spent on the hyenas.

One very interesting project that Gus conducted in conjunction with the computer specialist Peter Retief from Kruger Park concerned the closure of windmills. Over a four-year period various windmills were shut down and the movements of the ungulates in and around these closed windmills was monitored. Basically, what they found was that springbok movements were not affected, wildebeest moved out of areas where there was no water to areas that had water (in other words they were more water-dependent), and gemsbok moved into areas where the windmills had been closed to *brak* (dig) for minerals. If you watch gemsbok carefully at windmills, you will see that most of them do not drink any water but just eat ground containing minerals. In spite of this research little change in the water management policy of the area occurred. Gus had done his job in getting the results, and that was as far as he could go. One must remember that before watering points were put into the Kalahari in the early 1900s, all the animal species in the area had survived here for thousands of years without any permanent water. It is interesting to see that even today many tourists get upset when they find windmills that are not pumping water, believing them to be essential for the animals' survival.

At the beginning of 1981 Gus took three months of unpaid leave and we all went over to Scotland, where he spent quality scientific time with Hans Kruuk in Banchory and Martyn Gorman at Culterty

213

Field Station, with lots of very helpful interactions with other research students. This assisted him greatly with the analysis of the brown hyena data and the writing up of the results. We found it quite a shock to the system to go from Kalahari summer highs of over 38°C to the middle of a Scottish winter. It felt as if the boys and I spent most of the time huddled in front of the one and only heater in the cottage in which we stayed. Hans's and Martyn's wives were wonderful to us while Gus was busy and included us in a lot of what they were doing, introducing us to other local people with little ones. As we spent the last week touring beautiful Scotland, it wasn't all work for Gus.

It took us until the middle of 1981 to complete the first draft of the doctoral thesis, Gus writing it and me typing and drawing by hand all the diagrams and maps, using Letraset for the printing. Though it was extremely time-consuming, I enjoyed the work I had to do. I gave Gus a hard time, though, when he kept changing his mind about what went where, because it meant I had to retype many pages each time to get the changes to fit in (this was before word processors). I really admired Gus for his determination and tenacity in completing his thesis: it wasn't easy for him, particularly as he had no one at Nossob with whom to discuss his research. Collecting data is always the fun part. Sadly, many research scientists spend years collecting valuable information and then fail to write it up; as a result it is lost to science. Gus finally graduated towards the end of 1982.

It was about this time that we made the decision that we would leave the Kalahari before the children had to go to school. We could not face the thought of sending them to boarding school at the tender age of six. The nearest English school was in Kimberley, which was over 500 kilometres away. I was willing and felt able to home-school the boys but in those days one was not allowed to do so in the Northern Cape. In various discussions with the Education Department I was told that our children had to attend a registered school, and that the department would come and get them if I attempted home schooling. So Gus put out the word that within a couple of years he would like to have another job somewhere closer to schools.

CHAPTER 8

The
Final Years

OUR LAST TWO YEARS in the Kalahari turned out to be the most active social period we spent there. Mike Rosenberg from Partridge Films arrived from England in early 1982 during a very hot and dry period to investigate making a film on the Kalahari, with the well-known South African photographer Anthony Bannister. Gus was asked to write a background account of the important ecological features of the Kalahari as a basis for the film. It was called *Kalahari: Wilderness without Water*. Richard Goss and his partner Karen were Anthony's assistants, and Richard ended up taking a lot of the footage when Anthony was away. They all camped at Nossob and went through many of the difficulties we had experienced during our camping days. Like Doempie before me, I was able to rescue their belongings when we had heavy storms and they were not in camp. Though I always offered them the comfort of our home during particularly violent wind storms, they seemed to cope pretty well.

On a typically cold Kalahari night, Gus went out following spotted hyenas with Richard to do some filming. In the early hours of the morning the car battery went flat, and they were stuck. It is impossible to push a small car, never mind a Toyota Land Cruiser, in the Kalahari sand. They even tried jacking up a back wheel, winding a rope around it, and then all three pulling it as hard as they could in an attempt to kickstart the engine into life, but only managed to fall over each other in the process. Eventually they built a fire to keep warm and waited until 7 am to call me on the radio. Div kindly went out to rescue them.

A few months later David and Jenny Macdonald arrived from England to work on meerkats. David is the head of the very successful Wildlife Conservation Research Unit at Oxford University (WildCRU) and an incredibly prolific scientist. He and Gus had met when they were students under Hans Kruuk. We so enjoyed their company and had many intense discussions and lots of fun and laughter. Several years later Richard and David produced a wonderful wildlife documentary in the Kalahari on meerkats called *Meerkats United*.

A visiting German filmmaker also arrived at about this time to try to make a film on brown hyenas. At this stage Gus was unable to help him much as he had by then turned his attention to the spotteds. However, early one morning Gus found a brown hyena carrying a large piece of meat just south of Rooikop. That afternoon he and the filmmaker went out and, assisted by a tracker, found a den close by. Very soon three cubs of about five months and three a bit older made an appearance. They were all very tame and relaxed. Next, Phiri, whom we had marked as a cub in 1973, came to the den and suckled some of the cubs. The following day the whole family went back there with the German. He took photos of us all watching the den, and of Gus and the boys walking in the dunes and Michael whooping rather sweetly. This den was so accessible that we often went there with the boys in the late afternoons to watch the comings and goings of the hyenas.

Later in the year Hans Kruuk returned to Nossob, this time for a six weeks' stint to do some research with Gus on the little-known honey badger. It was a real struggle to catch any animals and they only managed to get a collar on one, after which it was shy and difficult to observe. However, aided by a tracker they managed to collect some data on its movements and also picked up a number of scats to see what they had been eating. Interestingly, some of them were full of scorpion remains. Although they only scratched the surface of knowledge of these weird and wonderful creatures, several years later Gus's student Colleen Begg and her husband Keith conducted a full study and, with the filmmaker David Hughes, produced another wonderful Kalahari film called *Snake Killers of the Kalahari*.

One morning at 5.30 after most of the tourists had departed from camp, we spotted a huge male lion near the entrance gate. Some of the staff joined me in the Kombi and we went to have a look at him. He crossed the road in front of us and then moved off behind the camp. Thinking that he was going to drink at the crib just outside the junior staff village on the way to the camp, I rushed off there to tell Michael's playmates to stay put. I had visions of them meeting up with the rather hungry-looking lion on their way to come and play. I then came back to camp to discover that a leopard had raided the ranger's chicken cage during the night, killing seven of them. There were feathers and bodies strewn all over the place.

When I got back to the house I found to my horror that I'd left the hot tap running in the kitchen sink with the plug in position. When first told about the lion I had rushed to the bathroom to pull the skewer out of the geyser (the only way we could get it to work) and had forgotten to turn off the tap. The kitchen, pantry and passage were flooded.

I then had to go to the office to prepare for radio session and a few minutes later my helper, Lassie, came rushing over carrying Stevie and a big thermos flask. He had somehow pushed his arm into the flask and couldn't get it out. It took a lot of coercing to calm him down enough to relax his muscles and extricate his limb. All in all, it was rather an interesting and eventful hour. It wasn't even 7 am and I felt quite exhausted.

In May 1983 Div and Alta were transferred to Mata Mata Camp. I had to take over running the camp at Nossob until a replacement was found, which only happened in August. Rian and Lorna Labuschagne were employed as the new ranger and camp manager. Lorna had a diploma in nature conservation and was not keen to be stuck in camp all the time. We tried to negotiate with the Parks Board to enable Lorna and I to share the job, but they were not willing to compromise. I felt very sorry for her. She loved the bush but had very little chance to experience it and was on her own for much of the time. Both she and Rian were wonderful with our boys, even looking after them on the odd night so that I could go out with Gus. We enjoyed their company. Michael frequently stayed with Lorna in the day when Stevie and I went out tracking with Hermanus and Gus. One such time we ended up getting sidetracked and followed a leopard spoor. It was fantastic checking in the sand how the animal had moved from bush to bush, had crawled along on its belly, and then had sprung onto its prey, a duiker. The sand is a story book in itself. Unfortunately the interpretation of spoor is the preserve of only a handful of Khomani San and other Kalahari locals these days. Sadly it is a dying art.

By this time Gus had been told he would be transferred to the Kruger Park to work on buffalo and elephant. His constant requests over the previous few years had paid off. We were thrilled that our boys' schooling was no longer an issue, but were very unhappy about the thought of leaving the Kalahari. Gus, too, was not sure whether forsaking meat-eaters for vegetarians was what he really wanted. We were to start in Skukuza at the beginning of 1984. Mike Knight was appointed as the new biologist in the Kalahari.

In view of the pending move, we packed up and moved out of our house by October, just keeping the bare necessities. Our home was now a family cottage in the camp. This was necessary so that renovations could start on the house. We fought hard for all sorts of improvements for the new occupants, especially for a larger pantry. Initially we were told that this was not possible, as no one in Skukuza had one. Never mind that we only went to town once every few months and had to buy in bulk. Fortunately we eventually won our case and the house was improved, making it much easier to live in. We were pleased for the Knights, but it made it more difficult for us to leave.

At the beginning of 1984 we were informed that we could not move to Skukuza as planned, as there was no accommodation for us. Mike and Anette Knight arrived in Nossob. Though Gus and I felt miserable now that everything was so final, we spent the next four months really enjoying our new neighbours and the Labuschagnes. It was a very happy time. We often ate together, took it in turns to go out with Gus, went game-viewing together and even started running in the evenings, Michael and Stevie also joining in for short distances. We would drive a couple of kilometres out of camp once the gates had closed for the tourists. One person would be designated driver and follow the runners and pick them up when they had had enough. I summed this time up when writing to the folks: 'It is so pleasurable doing things together and we have had more fun these last couple of months than we had in the 12 years we have been here.'

We then heard from Skukuza that Ian Whyte, who initially had been given the responsibility of carnivore research in Kruger, was actually more interested in looking at buffalo and elephant population dynamics instead of carrying on with his lion research. This meant that Gus would be able to work on carnivores. We couldn't believe our luck. For the sake of having the boys at home with us, Gus was prepared to work on anything, but that he was getting what he secretly wanted was too good to be true. We were thrilled.

Our departure from Nossob was fairly dramatic. The removal people finished packing all our belongings too late in the day to go through to Twee Rivieren before dark and had to spend the night at Nossob. The driver decided to turn the van around and park it near the exit gate for an early getaway the next morning. Unbeknown to any of us there was an old drain hidden under the sand just where he tried to turn and the truck fell into it. The driver was adamant that he was not going to unpack everything, so we tried pulling out the van with all the vehicles in camp, as well as levers to lift it, but it would not budge. Eventually we had to get the grader that was busy working near Union's End, 120 kilometres away, to come and pull it out. Our departure was delayed by two days.

When we eventually arrived in Skukuza in May, we discovered that our house still wasn't ready and we spent another three months in two different houses before we could unpack and get our lives straightened out. It was a long 10 months living out of boxes. But we were excited about the new adventure that we were about to embark on.

Our 12 years spent in the Kalahari was a privilege that we will never forget and will always appreciate. We learnt to make do when we ran out of specific foods and goods, and survived quite well without them. It made me realise that we clutter our lives with so many possessions and nonsense that are really not important and make life much more complicated than it needs to be. And to be able to become so involved with the hyenas was very special.

At our large Christmas family gatherings at Kasouga in the Eastern Cape, often with more than 40 of us, my nephews and nieces were taught from a tender age by their Uncle Gus about the wonders of hyenas. The saying 'Hyenas are beautiful' was drummed into them from the word go and later became a password to many exciting happenings and treats. And Gus has always said he thinks I'm beautiful, so I have never had anything to worry about.

Epilogue

WE HOPE THAT THIS BOOK helps to improve the image of hyenas, that it creates an awareness of what interesting and important animals they are, worth conserving, not only for their intrinsic value, but because of their beauty and fascinating behaviour. We hope too that the reader has been able to share with us the thrills and fulfilment that we were privileged to enjoy as we unravelled some of the night secrets of the beautiful and mysterious Kalahari.

In 1984 we left the Kalahari with mixed feelings. We had had a most exciting and interesting time there and had experienced some wonderful moments. On the other hand the isolation of the place made it difficult to raise a family and living in such a small community has its problems. We were to spend 22 years in Kruger and not for one minute regretted having made the move. Margie's duties as a mother meant that she was unable to assist Gus in the field with his carnivore work on lions, cheetahs, wild dogs and spotted hyenas, but she carved out an important niche in Skukuza with her social work in neighbouring communities and in running the Skukuza Marathon Club and the very popular Skukuza Half Marathon. However, the Kalahari still holds a very special place in our hearts – so much so that in 2006, on retiring from SANParks, we returned to the Kalahari and are once again working as a team, this time with cheetahs.